RISK MANAGEMENT PLAN
RMP*eSUBMIT USERS' MANUAL

RISK MANAGEMENT PLANS UNDER THE CLEAN AIR ACT

Risk Management Plans (RMPs) must be fully updated and re-submitted at least once every five years. Most RMPs were submitted by the initial deadline of June 21, 1999, and were updated and re-submitted in 2004 and 2009, unless they were updated and re-submitted previously. EPA released RMP*eSubmit to facilitate secure online RMP resubmissions beginning in calendar year 2009.

WHAT'S NEW?

Since 2009, you have been able to submit your RMPs online via RMP*eSubmit. This software is on EPA's secure network, the Central Data Exchange (CDX), which manages thousands of data submissions from states and industry to a variety of EPA programs. Several updates to the software have occurred since 2009.

Reporting requirements are the same as in 2004. While the format of your reporting changed in 2009, the 2004 rule changes will remain in effect.

You can access your RMP online anytime. Owners and operators of facilities, and individuals they designate as Preparers of their RMPs, can access the facility RMP at any time online.

RMP*eSubmit replaced RMP*Submit 2004. EPA no longer supports the RMP*Submit 2004 software.

NAICS Codes update. The 2012 version of the North American Industry Classification System (NAICS) codes has been integrated into RMP*eSubmit.

WHERE TO GO FOR HELP

RMP Reporting Materials

EPA's Website: http://www.epa.gov/emergencies/content/rmp/index.htm includes the Risk Management Program rule, Off-Site Consequence Analysis specific guidance and calculator, the list of regulated substances, links to factsheets, guidance documents, industry-specific model plans, FAQs, this *RMP*eSubmit Users' Manual*, and other information.

RMP Contacts

Under the Clean Air Act, Section 112(r), states can choose to take delegation of the Chemical Accident Prevention Program. If they do, they become the Implementing Agency for that state. Contact your EPA Regional staff for assistance. We maintain their contact information on our Website: http://www.epa.gov/emergencies/content/regional.htm

RMP Regulatory Questions

Contact the Superfund, TRI, EPCRA, RMP & Oil Information Center (also known as the "Info Center") for your policy, regulatory compliance, and reporting requirements questions.

    800-424-9346   Toll Free
    703-412-9810   Metropolitan DC area and international calls

    Monday – Friday: 10:00 AM – 5:00 PM Eastern Time

    Closed on Federal Holidays

RMP*eSubmit Software Support

Contact the RMP Reporting Center for specific software questions about RMP*eSubmit.

    703-227-7650   All domestic and international calls

    Monday – Friday: 8:00 AM – 4:30 PM Eastern Time

    Closed on Federal Holidays

    RMPRC@epacdx.net

TABLE OF CONTENTS

APPENDICES

CHAPTER 1 GETTING STARTED

Introduction

The *RMP*eSubmit Users' Manual* provides assistance in preparing and submitting Risk Management Plans (RMPs). RMP*eSubmit application software is the Web-based free tool EPA developed to facilitate electronic submission and is designed to significantly reduce errors in submission through validations for data elements. If you are unable to submit your RMP electronically, contact the RMP Reporting Center: 703-227-7650.

RMP*eSubmit allows you to:

- View the current version of your RMP (if you have already reported an RMP)
- Create a new RMP online (if this is a first-time submission)
- Make corrections to, or create a complete resubmission of your RMP
- Identify and correct any errors in your RMP prior to submission
- View a copy of the record of your RMP
- Print a copy of your RMP
- Submit your RMP online
- Access help screens to assist you in completing your RMP

You must be registered as a Preparer and/or Certifier in the Central Data Exchange (CDX), RMP Program in order to use RMP*eSubmit.

> **Definitions:**
>
> **Central Data Exchange (CDX)** is a secure, online location on EPA's network. CDX provides standardized and secure information collection services and infrastructure for EPA program partners. For example, CDX manages several Agency regulatory and monitoring programs, receiving submissions from facilities.
>
> **Certifiers** are facility owners or operators who must certify the accuracy and completeness of the information reported in the RMP. They have signed and submitted a one-time Electronic Signature Agreement (ESA) to the EPA. The ESA legally binds the Certifier's electronic submission to their signature. Only Certifiers can submit the RMP.
>
> **Preparers** have been granted permission by a facility to access the facility's existing RMP. They prepare data for a new or updated RMP. CDX notifies the Certifier when the new or corrected RMP is ready for the Certifier's review and submission. Preparers cannot submit the RMP.

Before You Start

Are you subject to the RMP reporting requirements? Check the requirements on EPA's Web site: http://www.epa.gov/emergencies/content/rmp/index.htm, or call the appropriate contact on "Where to Go for Help" page at the beginning of this manual. For updates and resubmissions, check your 5-year anniversary date if you already have an RMP in the system.

What is your EPA Facility ID? If you already have an RMP in the system, you have an EPA Facility ID. It is essential that we match your new RMP to any earlier versions. Your EPA Facility ID appears in a letter sent by the RMP Reporting Center after your first-time submission. The EPA Facility ID has twelve digits. Call the RMP Reporting Center if you need help: 703-227-7650.

What category are your processes in: Program Level 1, 2 or 3? Each process at your facility having more than a specified amount (threshold quantity) of a covered chemical (regulated substance) will be in one of these categories. The category determines some of your reporting requirements and governs how you enter the data. More about Program Levels can be found later in this chapter.

Is your RMP a First-time Submission, a Resubmission, or a Correction?

> **Definitions:**
>
> A **First-time Submission** means that an RMP has never before been submitted for your facility (by you or any previous owner/operator). This requires that you enter information for all nine sections of the RMP. Elements for all nine sections are discussed in Chapter 2.
>
> A **Resubmission** is an update of all nine sections of your RMP. If you are the owner or operator of an RMP-covered facility, EPA's Chemical Accident Prevention regulations at 40 CFR part 68 require that you fully update and resubmit your RMP at least once every 5 years. Resubmitting your RMP will re-set your five-year anniversary date. If you have previously submitted an RMP, your facility information will be pre-populated in RMP*eSubmit.
>
> A **Correction** should be used to report administrative or other changes at your facility (e.g., changes in emergency contact information, facility address, or change in accident history). This does not require an update of your entire RMP. Submitting a corrected RMP does not change your five-year anniversary date. If your facility has not resubmitted an RMP by its anniversary date, you will not be able to make an RMP Correction and will be required to resubmit your RMP.

CDX and the Registration Process for RMP*eSubmit

The RMP*eSubmit application allows you to securely submit your RMP over the Internet through the EPA Central Data Exchange (CDX).

The requirements to submit using RMP*eSubmit are:

1. You must have Internet access
2. The Certifying Official (Certifier) and Preparer must have Central Data Exchange (CDX) accounts (see the "Registration" section to learn how to obtain those accounts)
3. The Certifier must complete an Electronic Signature Agreement (ESA) to obtain the Authorization Code (AuthCode) for preparing a submission. The Authorization Code is necessary to prepare and submit an RMP
4. Preparers must activate their RMP*eSubmit access in CDX using the unique facility Authorization Code provided to them by the Certifying Official for their facility.

Getting Started with RMP*eSubmit

To use RMP*eSubmit, the Certifying Official for your facility (typically the facility owner or operator or a designated senior management official employed by the facility) must first be registered in CDX. Additionally, the Certifying Official must complete and sign an Electronic Signature Agreement (ESA) which will be used to verify the identity of the Certifying Official in the system. The ESA must be mailed to the RMP Reporting Center and approved prior to the Preparer being able to prepare or the Certifier being able to certify an RMP using RMP*eSubmit.

Registration for RMP*eSubmit

The registration process for the RMP*eSubmit application is different if you already have a CDX user account (such as for previously registering to use TRI-MEweb or any other CDX application), and is based on whether your role is a Certifying Official or a Preparer.

The following information will take you step-by-step through the CDX registration process and will help you get started as a Certifier or Preparer.

Registering for CDX User Account

A CDX account is needed to use the RMP*eSubmit application. To register for an account, log into CDX at http://cdx.epa.gov, read the *Warning Notice and Privacy Policy,* then click "Register with CDX."

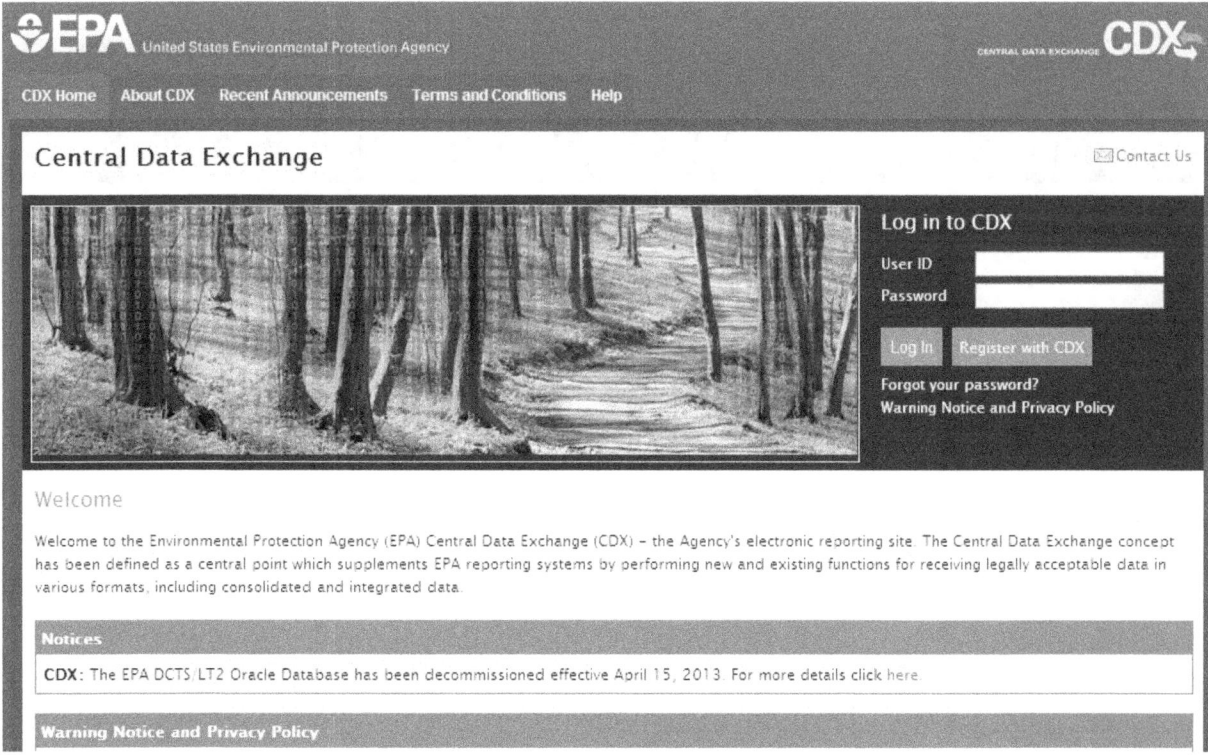

Read the *Terms and Conditions.* Then choose "I Accept" and click "Proceed".

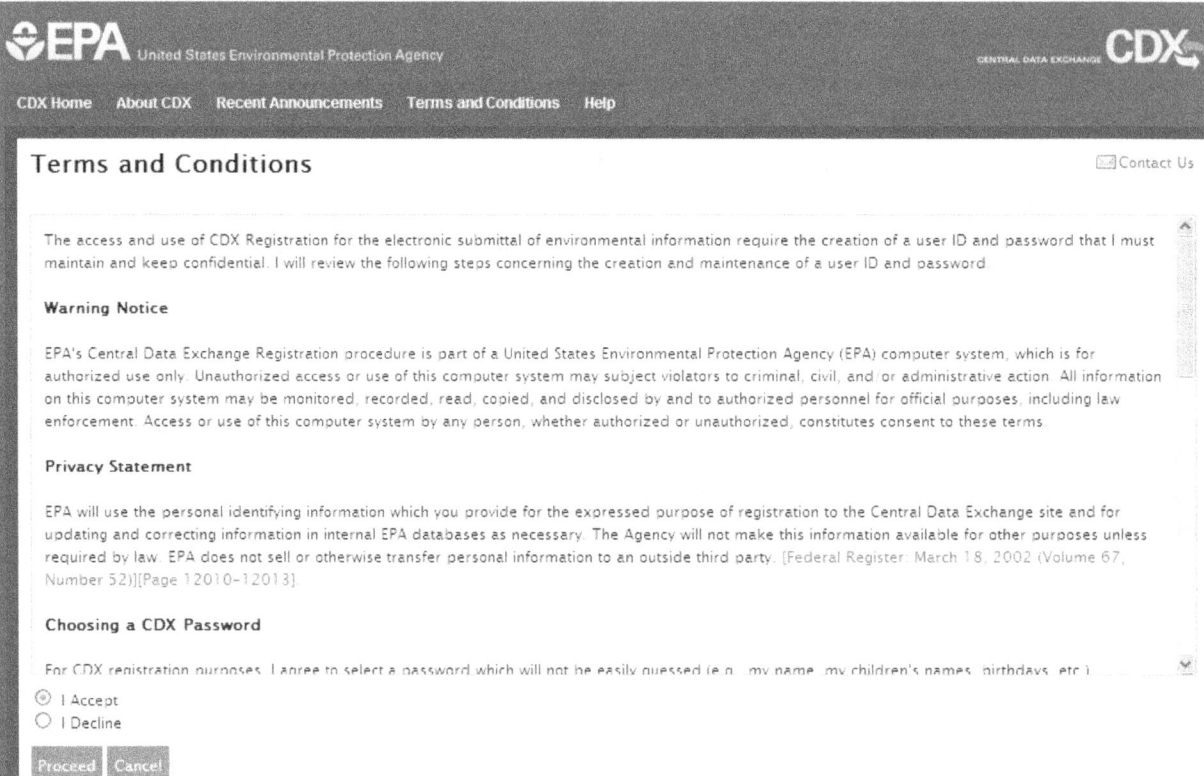

The *Core CDX Registration* page contains the Active Program Services List. Scroll to the "RMPESUBMIT: Risk Management Plan" link and click.

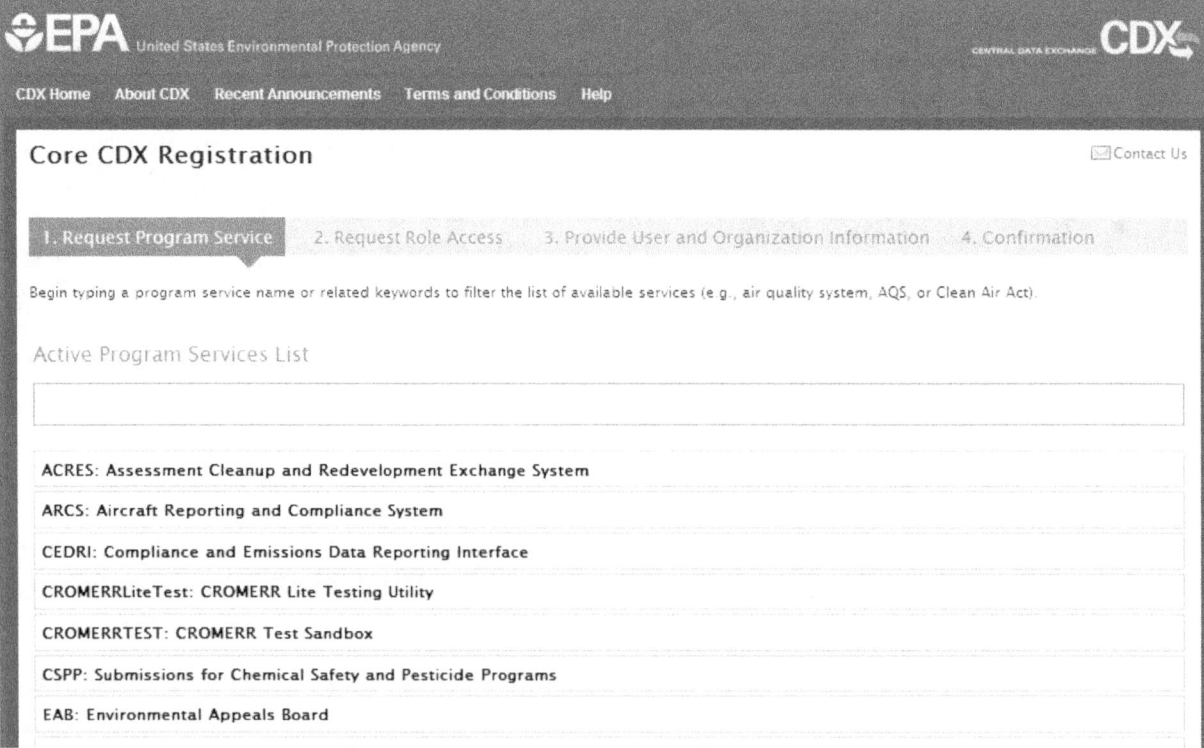

Using the drop-down menu, select "Certifying Official" and click the "Request Role Access" button.

> *NOTE: To add only the "Preparer" role to your CDX account, you will need to visit*
> *https://cdx.epa.gov/ and select "Register with CDX". Read the Terms and Conditions*
> *thoroughly. Chose the "I Accept" radio button at the bottom of the page, then click the*
> *"Proceed" button. (To add the Preparer role, go to page 18.)*

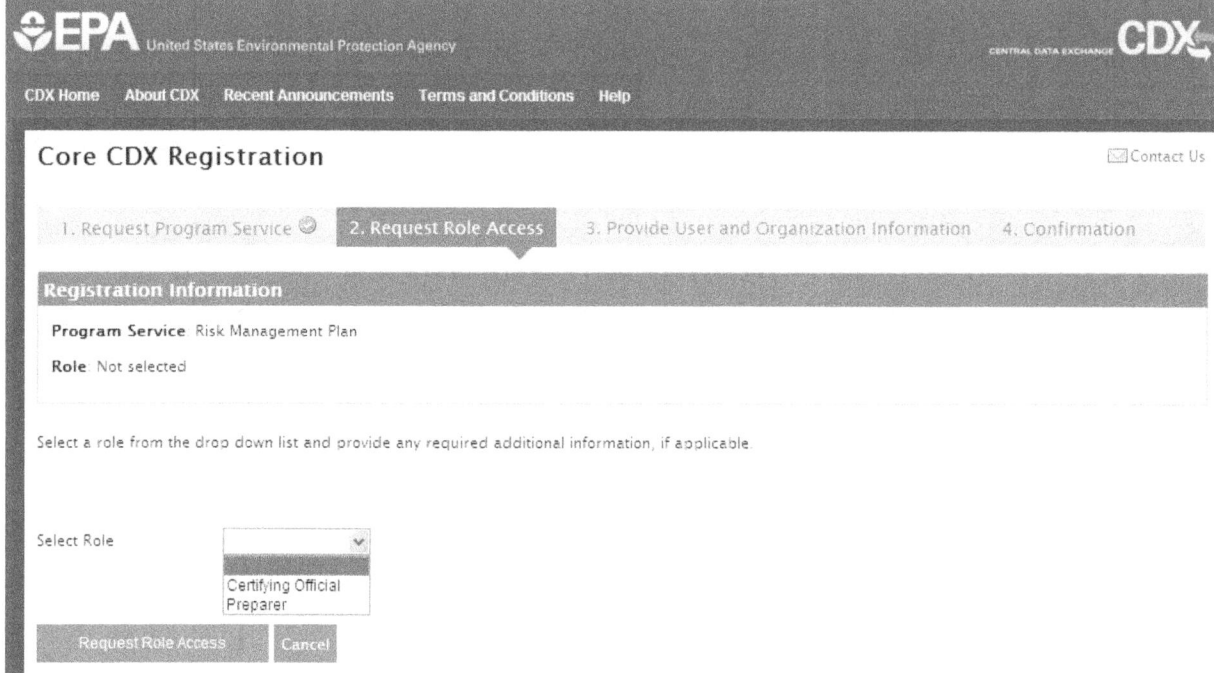

This brings us to the *Core CDX Registration* page. Scroll down to "Part 1: User Information" section and provide the requested information.

Once completed, move on to "Part 2: Organization Information".

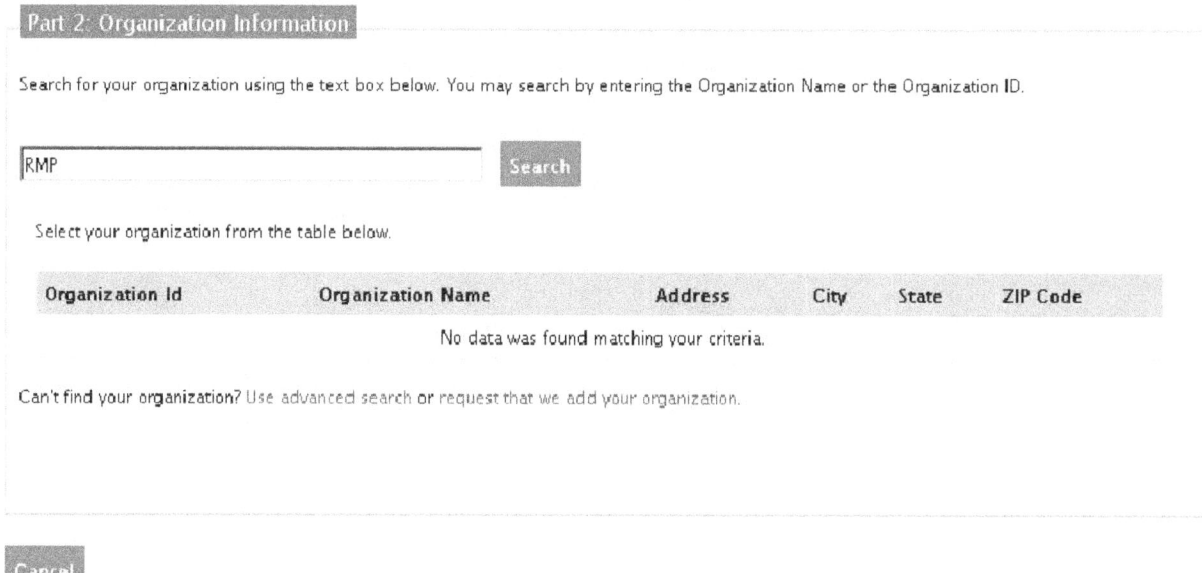

Use the search function to find your organization or you can request to add your organization. To add your organization, click on "request that we add your organization" link. Then, enter your organization information and select "Submit Request for Access".

Find your organization by using the search function either by "Organization Name" or "Organization ID". Click on your organization's corresponding "Organization ID" link.

Complete the additional contact information and select "Submit Request for Access".

You will be directed to a *Confirmation* page. Shortly afterward you will receive an email confirmation.

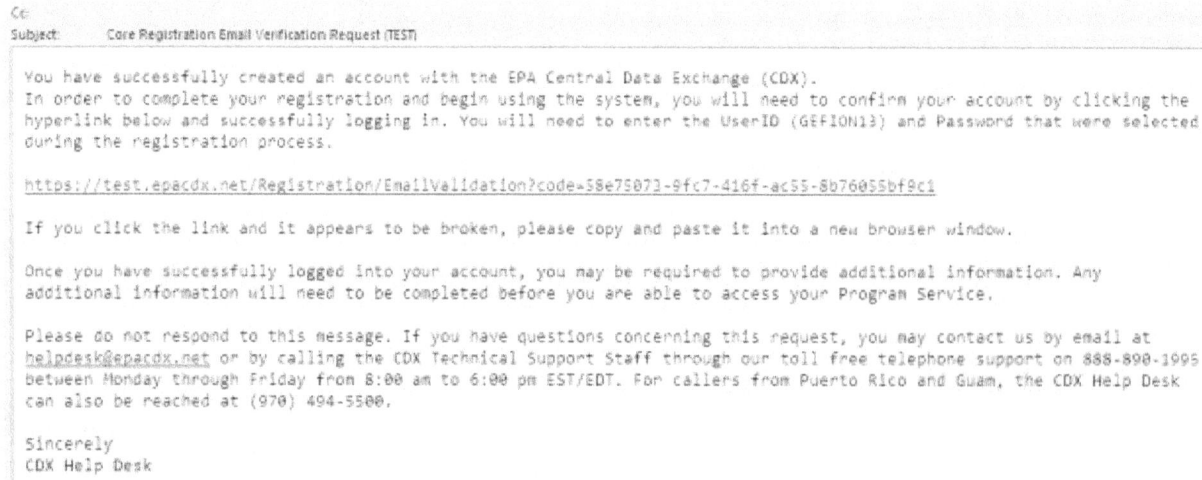

Follow the attached link in the email to activate your account.

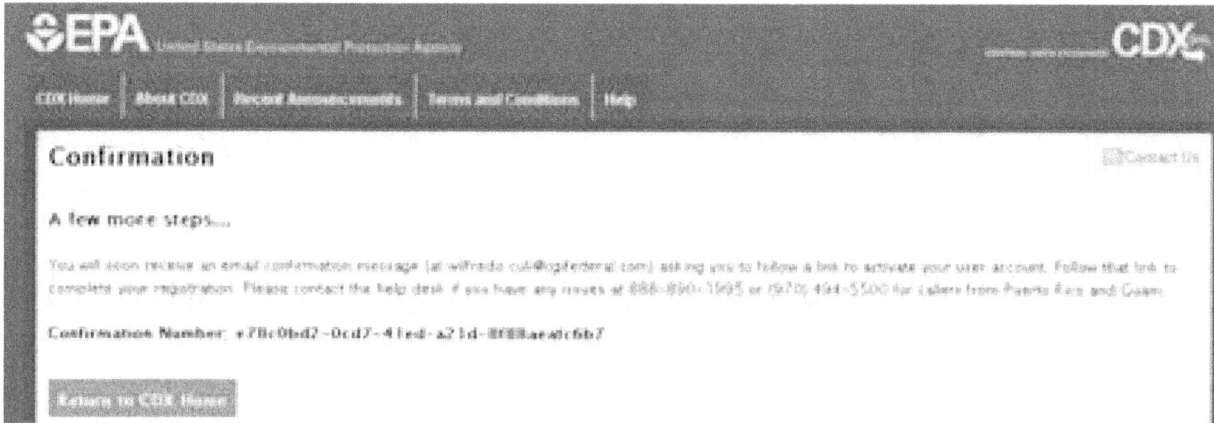

Once you've clicked the link provided in your email, you'll be directed to the *Central Data Exchange* page. Log in using your CDX User ID and Password.

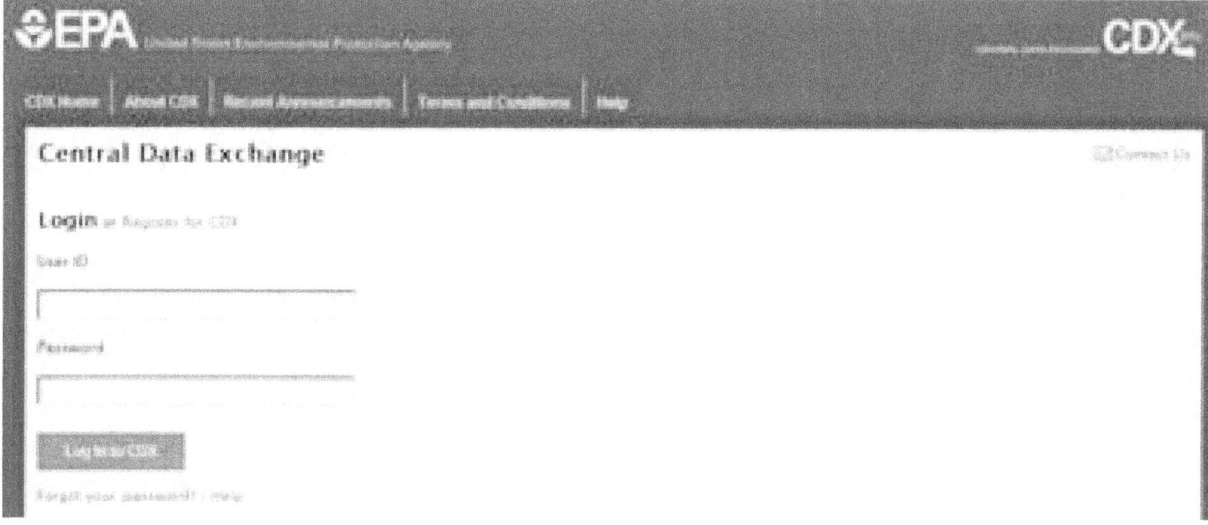

Select and answer five Challenge Questions and click the "Save Answers" button.

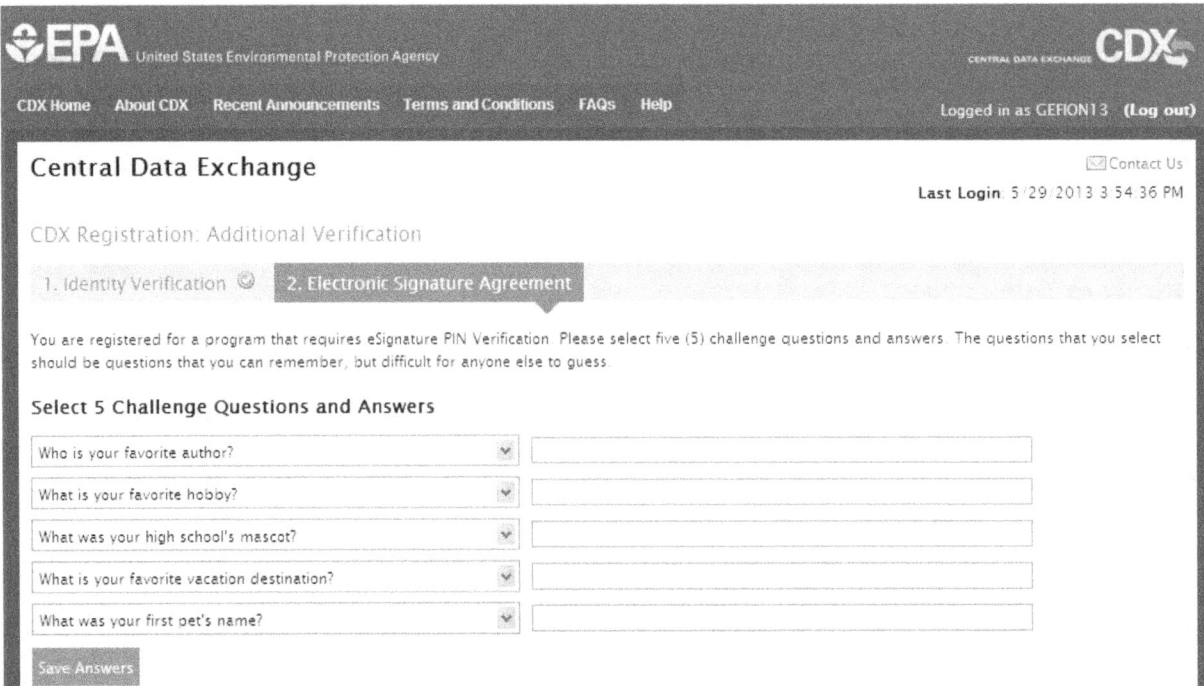

This takes us to the *Central Data Exchange* page. Look for the *MyCDX* tab. Click the "Certify Submission" link to add RMP facilities.

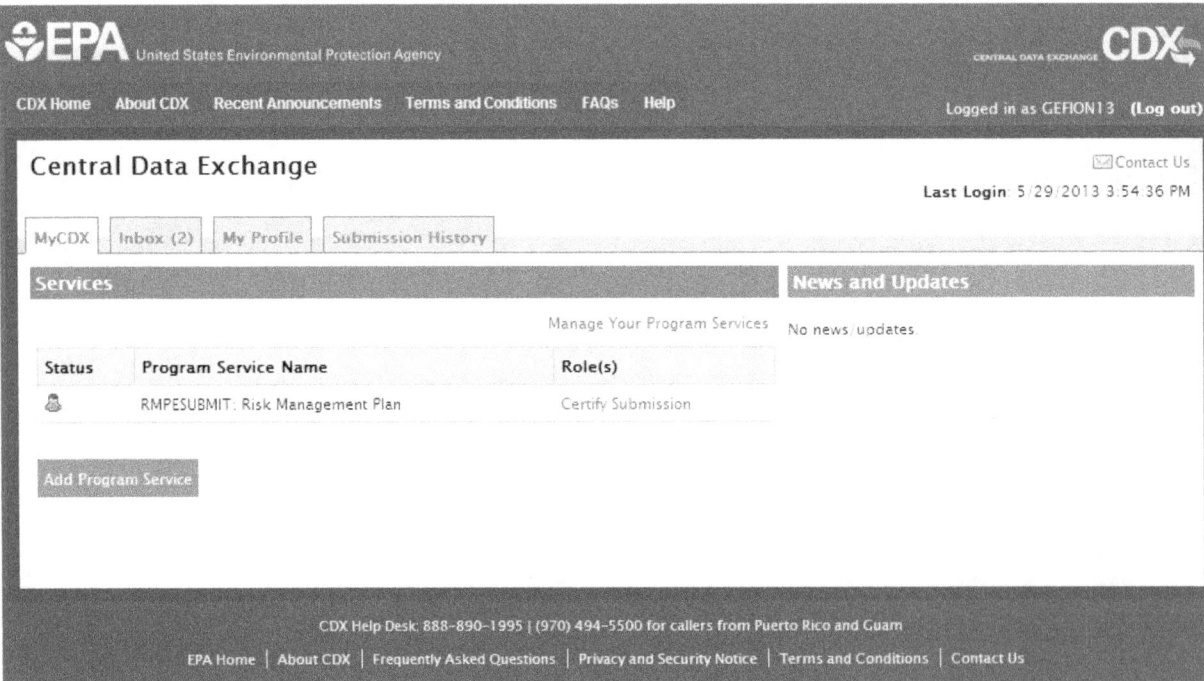

Click the "Add New RMP Facilities" link to create your Electronic Signature Agreement (ESA) for each facility you will be certifying.

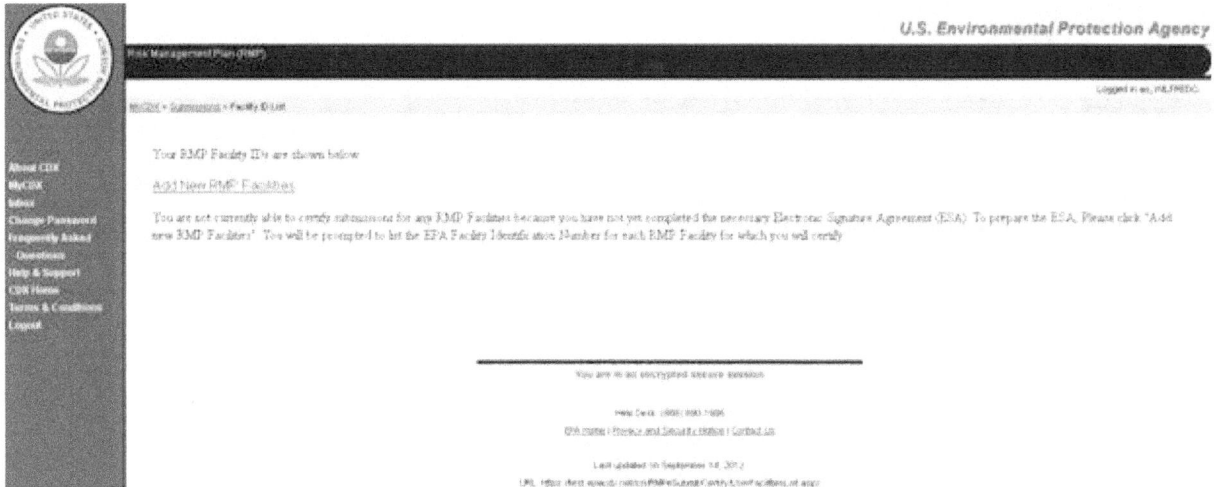

Now add the facility you want to be associated with your profile. To add more than one facility, click the "Add another Facility" button, then click "Save".

It is important to enter the correct Facility ID of the specific facility for which you'll be submitting an RMP. For a new facility/first-time submittal, the Facility ID field can be left blank, but you need to furnish the complete facility information (name and full location address). Once you click the "Save" button, the ESA (with relevant facility information) will appear.

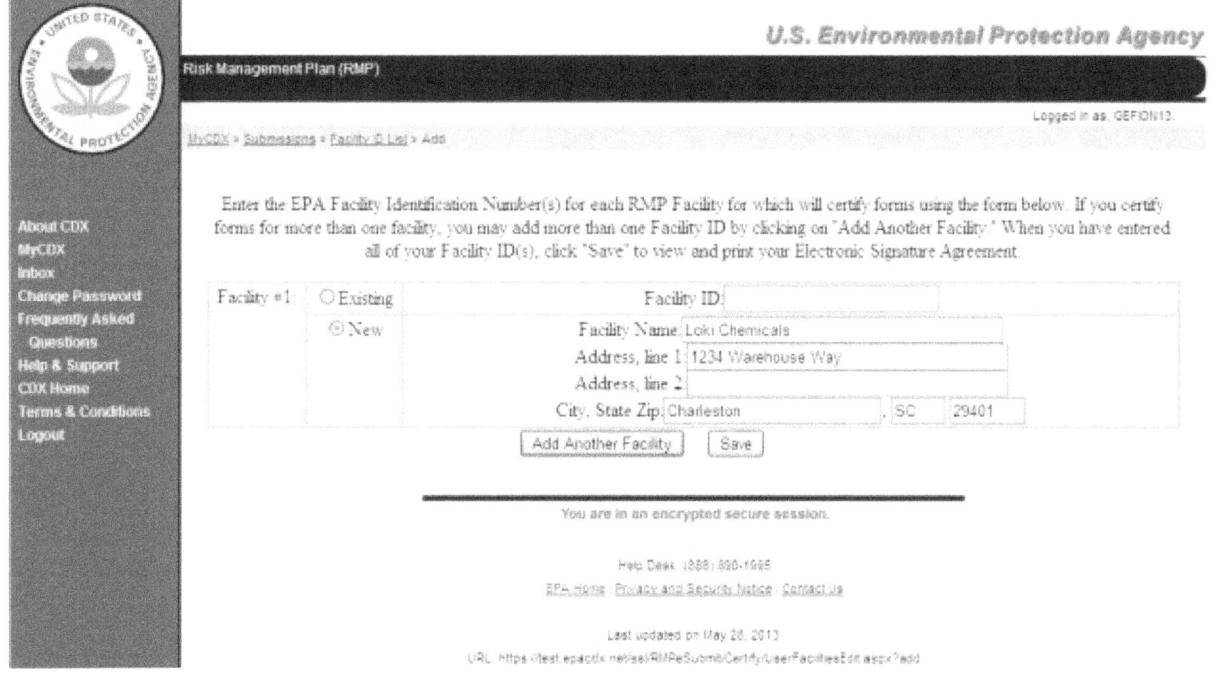

Upon, completion, CDX will prompt you to print the ESA. <u>Print **all** ESA pages</u>.

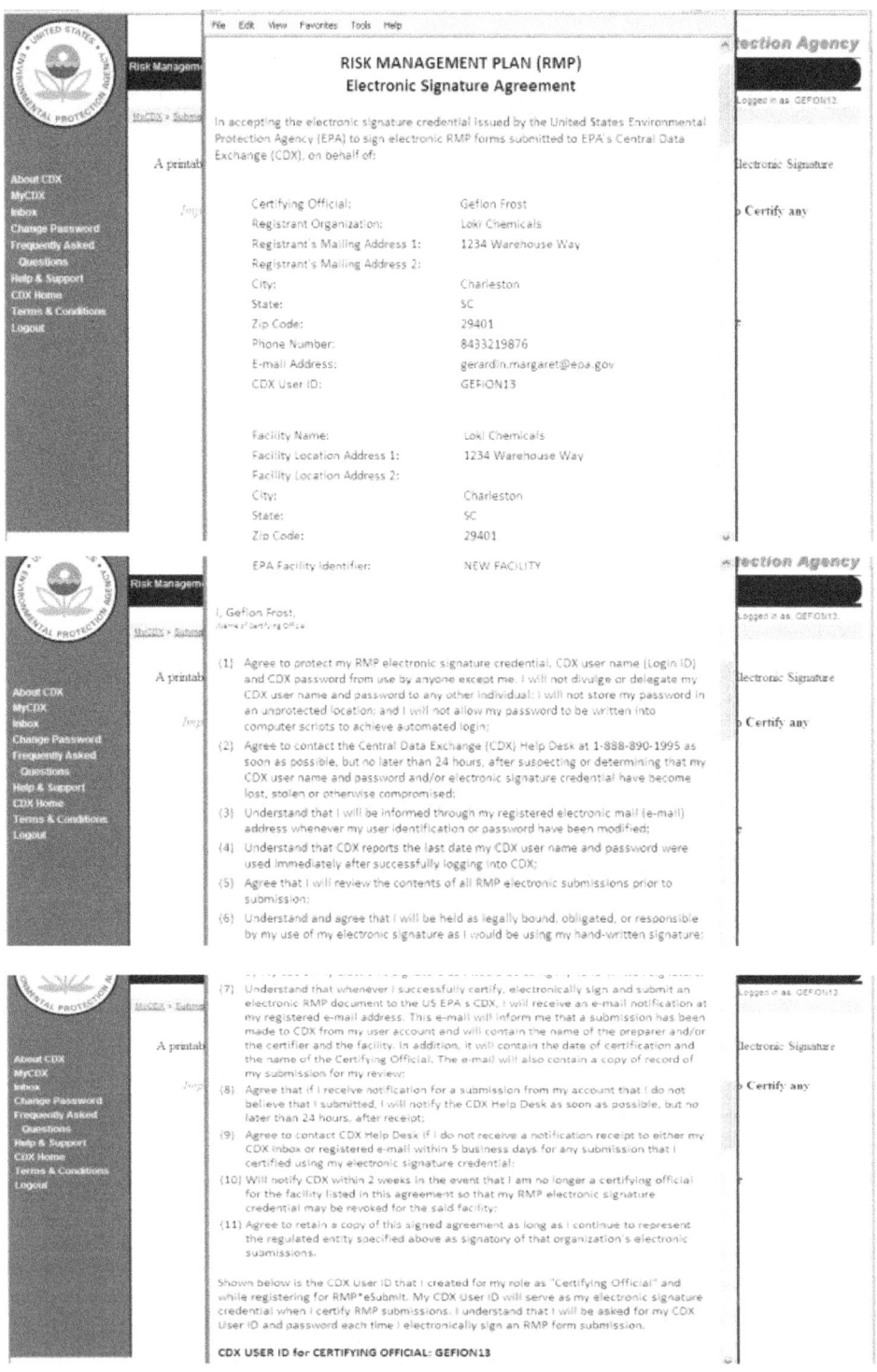

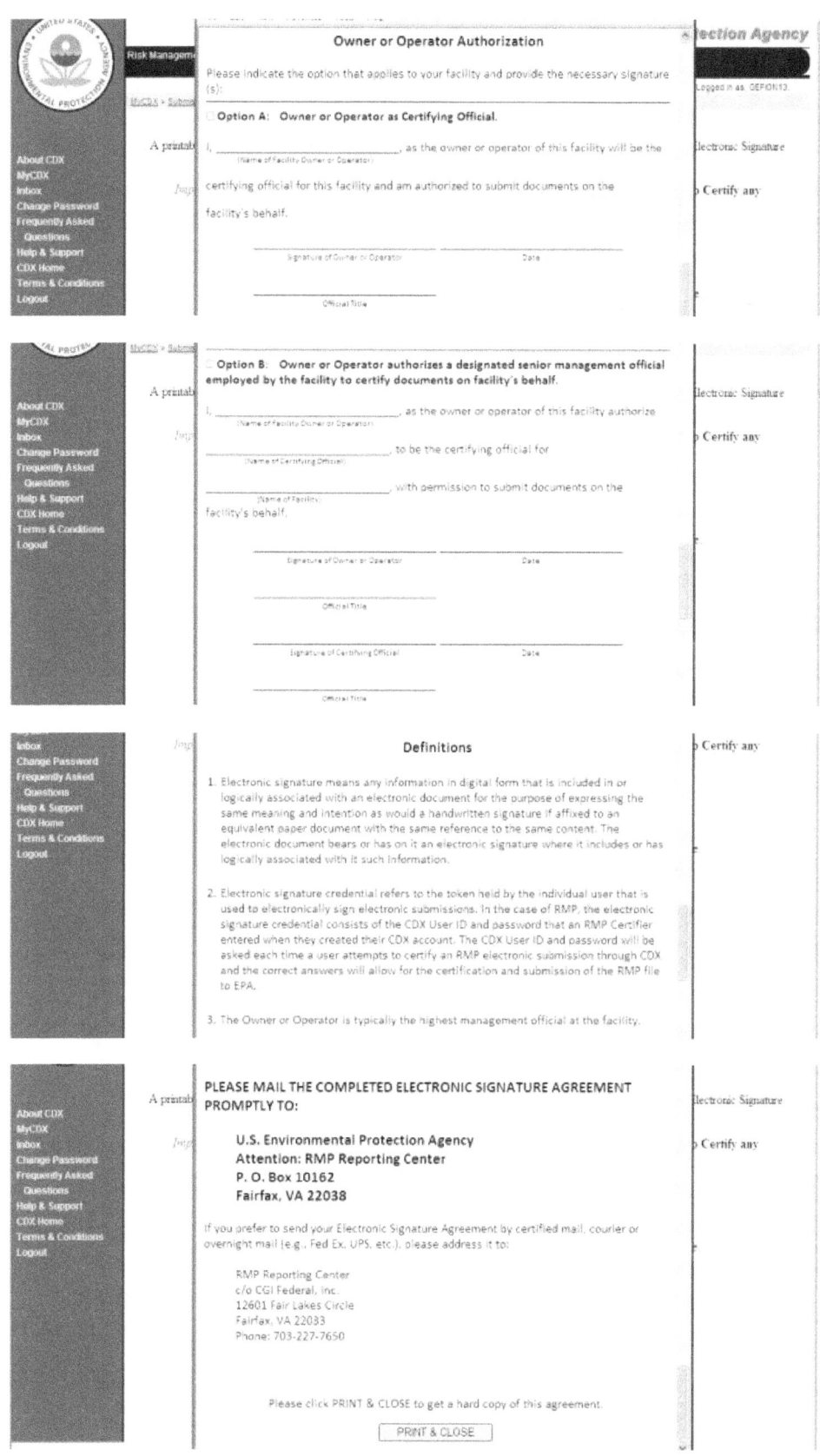

Owner or Operator Authorization

Please indicate the option that applies to your facility and provide the necessary signature(s):

Option A: Owner or Operator as Certifying Official.

I, _____, as the owner or operator of this facility will be the
(Name of Facility Owner or Operator)

certifying official for this facility and am authorized to submit documents on the
facility's behalf.

_____   _____
Signature of Owner or Operator   Date

_____
Official Title

**Option B: Owner or Operator authorizes a designated senior management official
employed by the facility to certify documents on facility's behalf.**

I, _____, as the owner or operator of this facility authorize
(Name of Facility Owner or Operator)

_____ to be the certifying official for
(Name of Certifying Official)

_____, with permission to submit documents on the
(Name of Facility)

facility's behalf.

_____   _____
Signature of Owner or Operator   Date

_____
Official Title

_____   _____
Signature of Certifying Official   Date

_____
Official Title

Definitions

1. Electronic signature means any information in digital form that is included in or
   logically associated with an electronic document for the purpose of expressing the
   same meaning and intention as would a handwritten signature if affixed to an
   equivalent paper document with the same reference to the same content. The
   electronic document bears or has on it an electronic signature where it includes or has
   logically associated with it such information.

2. Electronic signature credential refers to the token held by the individual user that is
   used to electronically sign electronic submissions. In the case of RMP, the electronic
   signature credential consists of the CDX User ID and password that an RMP Certifier
   entered when they created their CDX account. The CDX User ID and password will be
   asked each time a user attempts to certify an RMP electronic submission through CDX
   and the correct answers will allow for the certification and submission of the RMP file
   to EPA.

3. The Owner or Operator is typically the highest management official at the facility.

**PLEASE MAIL THE COMPLETED ELECTRONIC SIGNATURE AGREEMENT
PROMPTLY TO:**

**U.S. Environmental Protection Agency
Attention: RMP Reporting Center
P. O. Box 10162
Fairfax, VA 22038**

If you prefer to send your Electronic Signature Agreement by certified mail, courier or
overnight mail (e.g., Fed Ex, UPS, etc.), please address it to:

RMP Reporting Center
c/o CGI Federal, Inc.
12601 Fair Lakes Circle
Fairfax, VA 22033
Phone: 703-227-7650

Please click PRINT & CLOSE to get a hard copy of this agreement.

[PRINT & CLOSE]

Once you have printed **all** pages of the ESA and closed the document, a *Successfully Registered* confirmation page will appear. This completes your registration as a Certifier.

Next, follow the instructions on how to sign (Wet Ink Signatures). Mail **all** pages of the completed ESA to the RMP Reporting Center. The RMP Reporting Center will verify the ESA for completeness. It will then approve and generate the Authorization Code (AuthCode) needed to prepare an RMP for each facility. Since the ESA must be mailed, received, and processed, please allow one week from the date of receipt for your ESA approval.

An email containing an AuthCode for each requested and approved facility will be sent to the Certifying Official at the email address provided during CDX Registration.

```
Cc:
Subject:     Electronic Signature Agreement (ESA) (TEST)

*******************************************
This was sent from the TEST environment!
*******************************************


Your Electronic Signature Agreement (ESA) has been received and APPROVED for the facility/facilities shown below.
Following are the Authorization Codes assigned to each facility that you have requested access for.

Please provide the appropriate Authorization Code to the preparer that you assign for each facility/facilities. These
Authorization Codes are unique and are associated to each Facility (EPA Facility ID). Preparers must use the correct
Authorization Code for each facility to prepare a submission using RMP*eSubmit. For more information please follow the
instructions in the RMP*eSubmit  - user guidelines.


Authorization Code: 9df89f39-6e42-4436-9341-80f8bd1391e6


EPA Facility ID: 1000 0021 5689


Facility: Loki Chemicals
1234 Warehouse Way
Charleston SC, 29401


United States Environmental Protection Agency Central Data Exchange
```

Once the email has been received, the AuthCode will be used to prepare RMP. The Certifying Official is responsible for providing the AuthCode to the Preparer so that they can register in CDX and prepare the RMP.

> NOTE: For first-time submissions, the email to the Certifier will include a new Facility ID in addition to the AuthCode. Once the Preparer enters the AuthCode, they will be directed to the RMP*eSubmit welcome screen with the new ID and location address.

> NOTE to Preparers: The AuthCode is a long string of letters and numbers. Please copy the AuthCode from your email and paste it in the box when prompted.

> NOTE: If you already have the Certifier Role and you also want to be a Preparer, you can access this screen by returning to MyCDX. Click the "CDX Home" link on the top left hand side and click "Go to MyCDX". On the MyCDX page, choose "Manage Your Program Services". Next, click "Request a New Role" under the "RMPESUBMIT: Risk Management Plan" link.

Select "Preparer" from the drop-down menu and click "Add selected role". Enter your AuthCode and click the "Save" button.

Your *MyCDX* page will now display the roles that you have added to your profile (i.e., For Certifying Official – Certifying Submission, and for Preparer – Prepare Submission).

*NOTE: A Preparer's MyCDX page will only have the "RMP*Prepare Submission" link.*

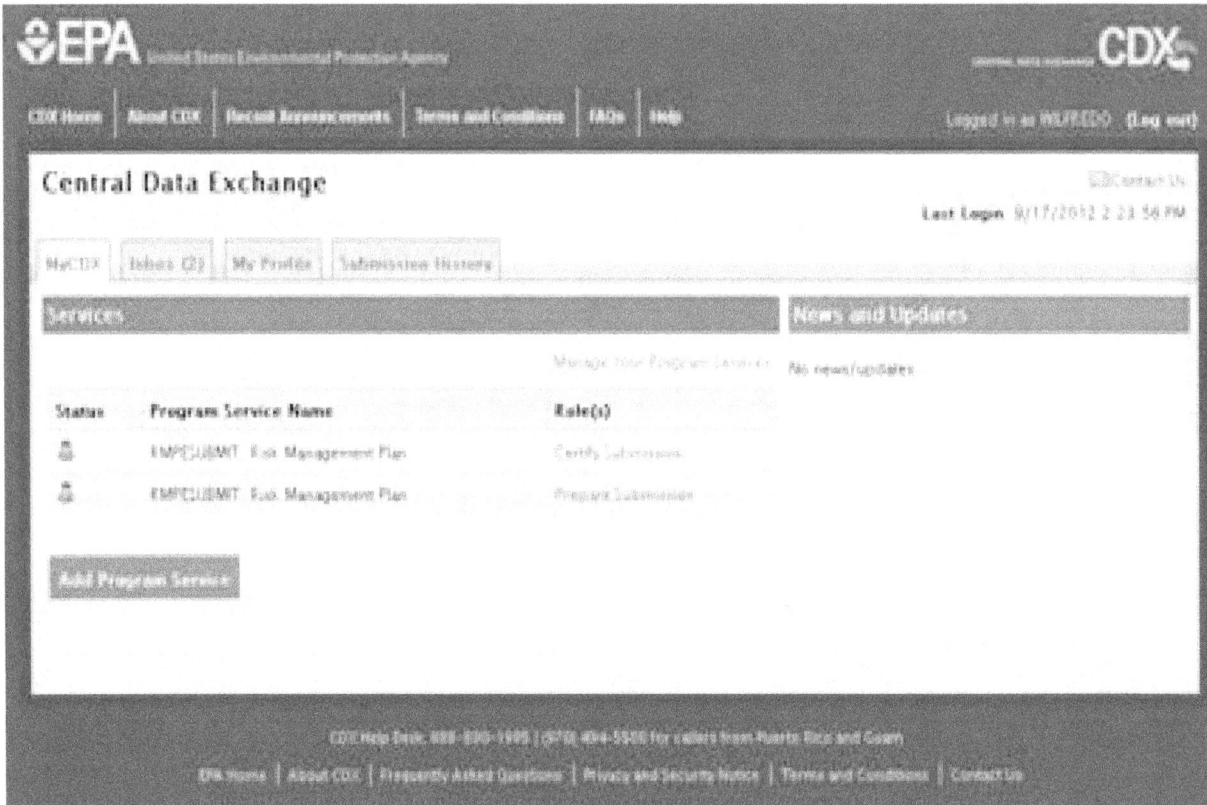

Click the "Prepare Submission" link and you'll be able to select the Facility ID you need to update the RMP. (If you are assigned multiple RMPs, a drop-down list of the Facility IDs assigned to you will be shown. Choose the correct Facility ID and Proceed.)

Preparing RMP and Facility ID Option Page

Once you select a facility ID from the drop-down list, click "Proceed" and you'll be directed to the *RMP*eSubmit Home* page.

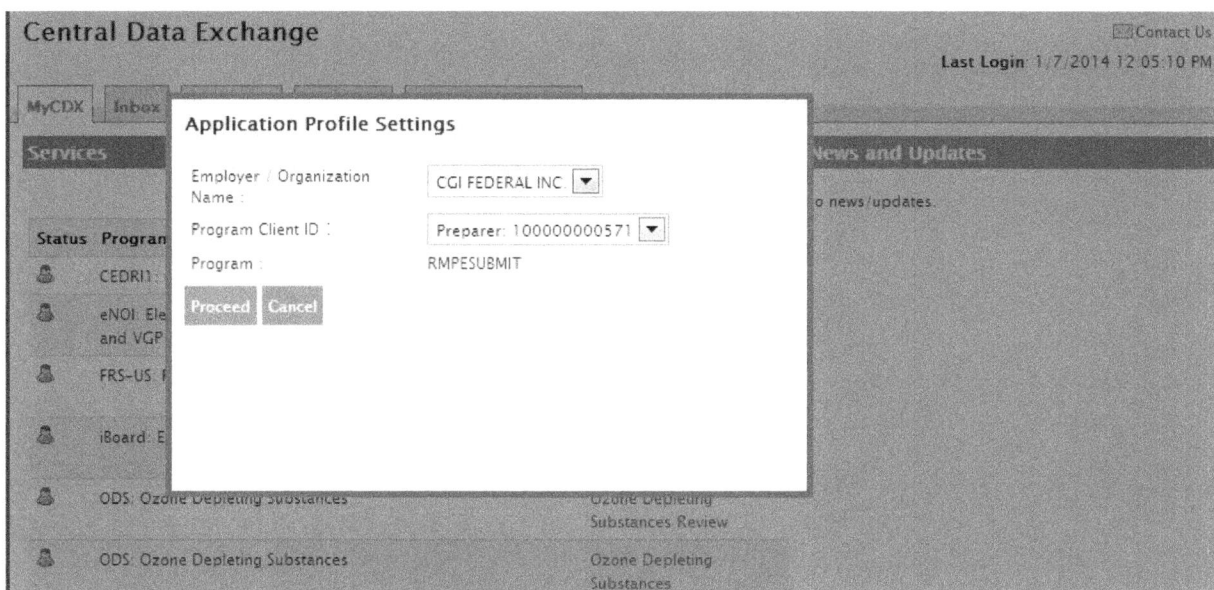

From the *RMP*eSubmit Home* page you'll prepare your facility's Risk Management Plan (First-time Submission, Correction, or Resubmission) using the **RMP*eSubmit** application. Once completed, you'll submit the RMP to your Certifying Official for certification.

> *NOTE to Preparer: If you are the Preparer for one facility, you'll directly be accessing the RMP*eSubmit Home page. If you are the Preparer for more than one facility, you'll see a dropdown list from which to choose the facility that needs to be updated (as shown in the previous screen shot).*

For a first-time submission, the *Home* page will resemble the screenshot below.

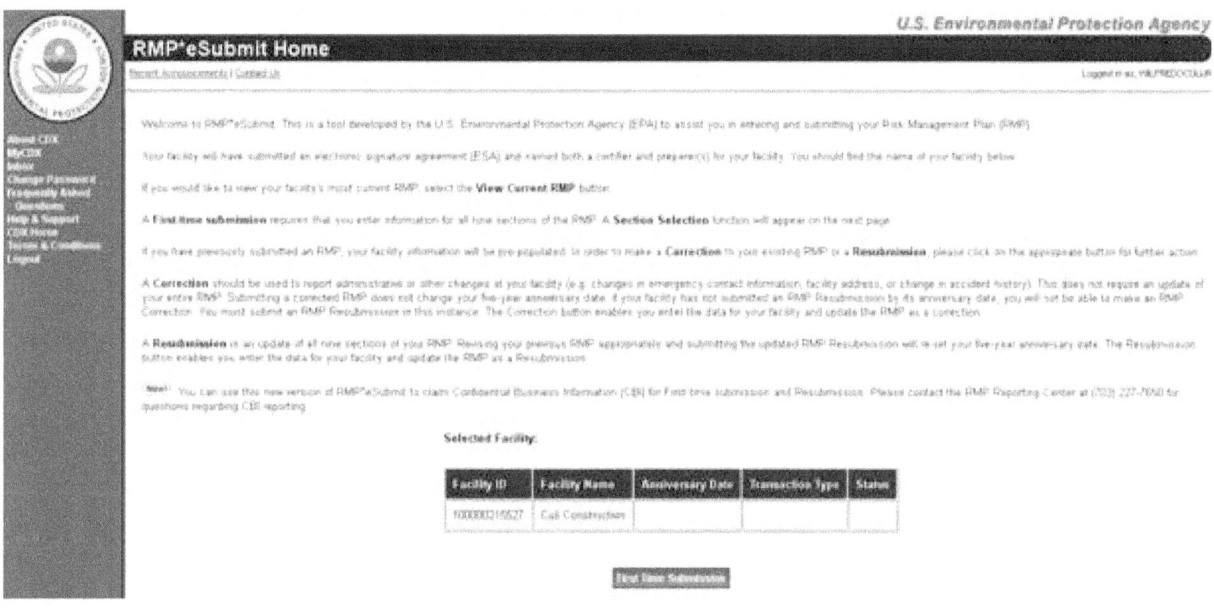

For a Correction or Resubmission, the screenshot will resemble the screenshot below.

Adding the Preparer Role

Select the appropriate button (Correction or Resubmission) and you will have access to the RMP*eSubmit program to update your RMP.

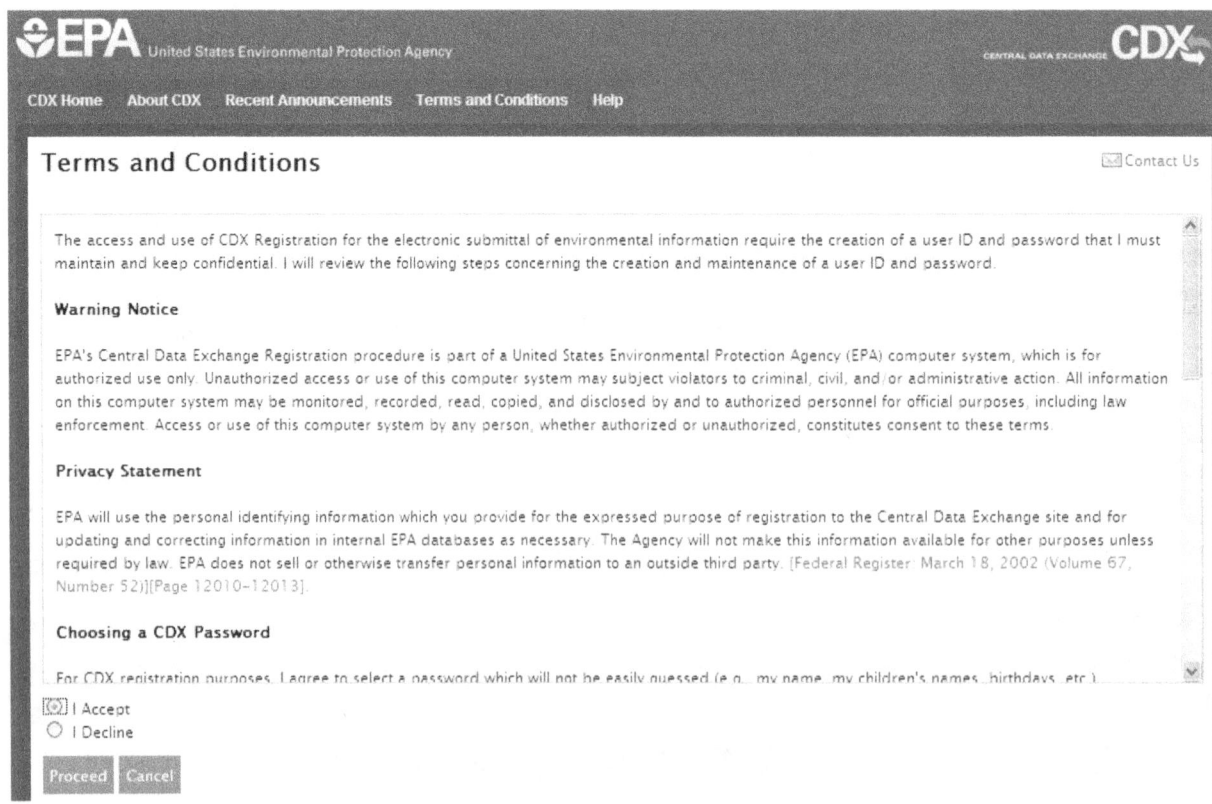

On the *Core CDX Registration* page, scroll to the "RMPESUBMIT: Risk Management Plan" link and click.

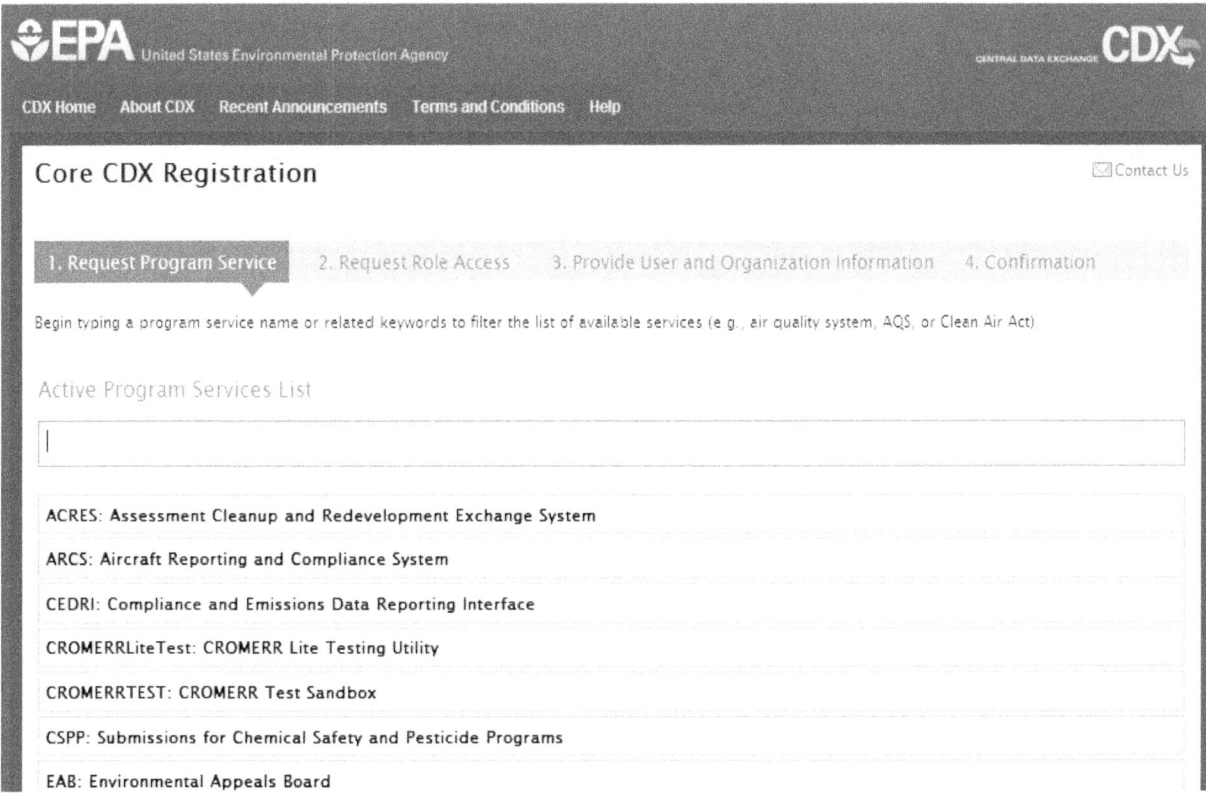

This brings us to the *Request Role Access* section. This time, let's choose "Preparer" and click the "Request Role Access" button.

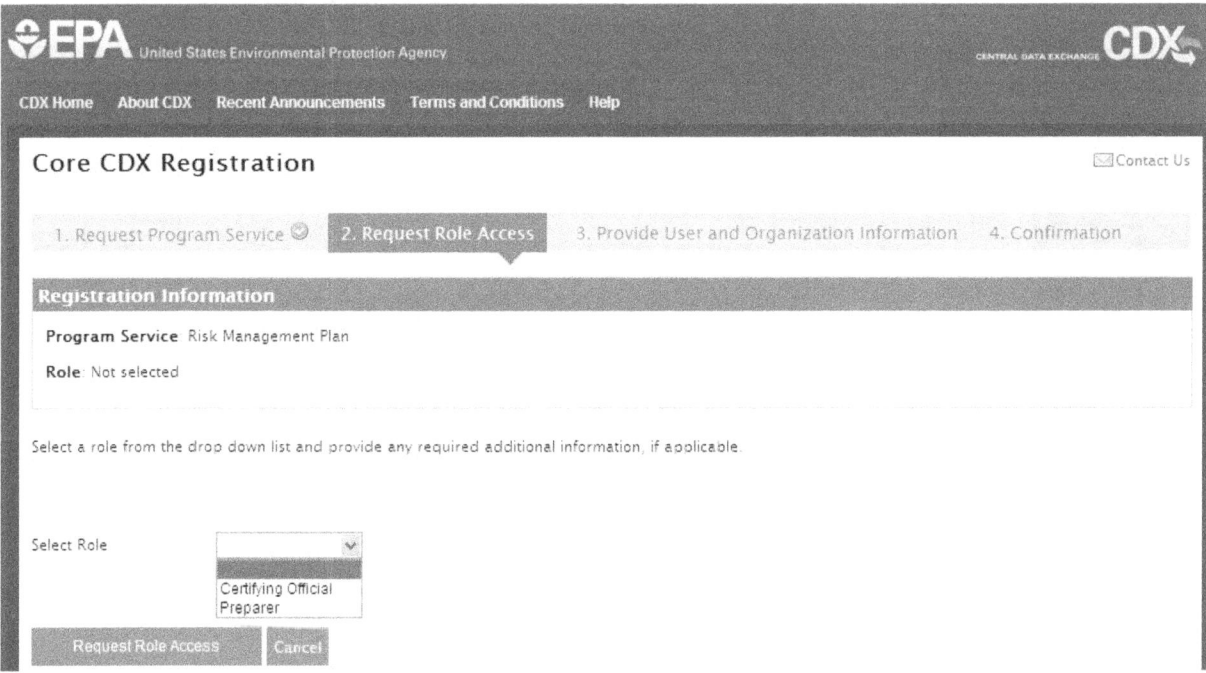

Enter (paste) the AuthCode in the box provided and click "Save". The AuthCode will be provided by your Certifying Official.

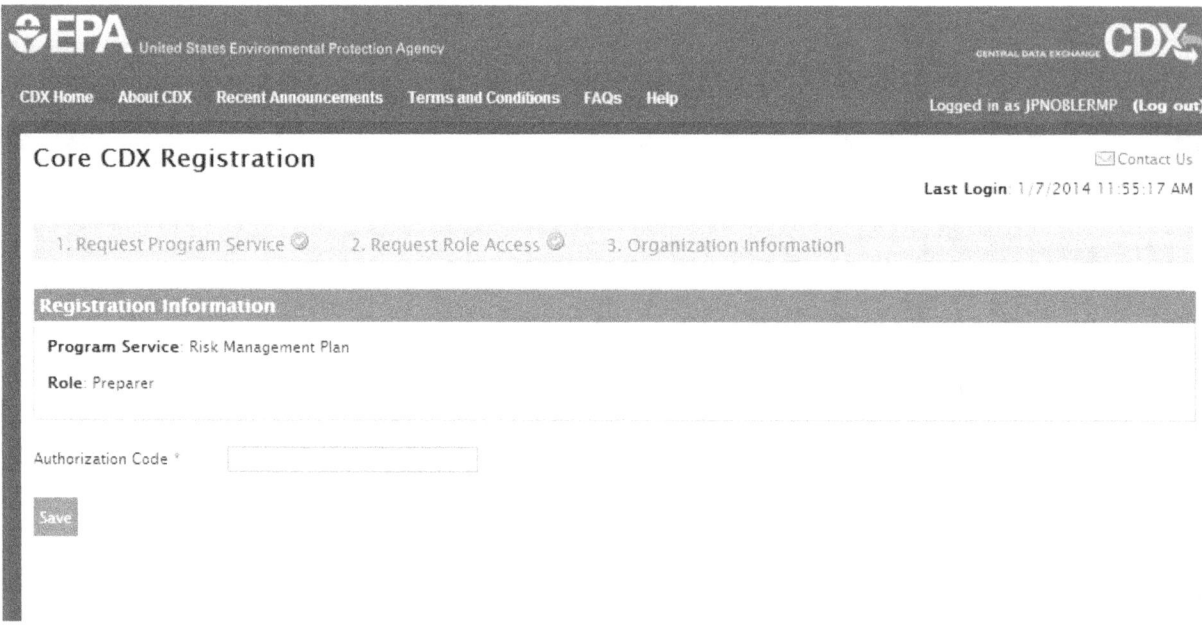

For a First-time submission, the *Welcome* page will resemble the screenshot below.

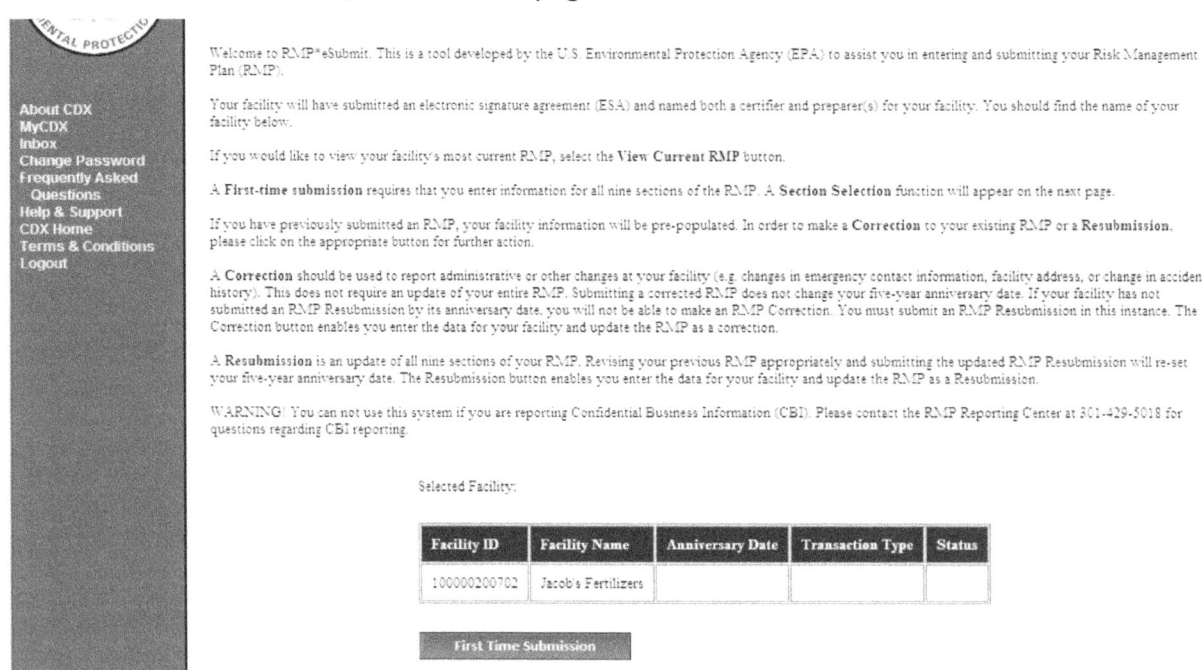

For a Correction or Resubmission, the screen will resemble the screenshot below.

Welcome to RMP*eSubmit. This is a tool developed by the U.S. Environmental Protection Agency (EPA) to assist you in entering and submitting your Risk Management Plan (RMP).

Your facility will have submitted an electronic signature agreement (ESA) and named both a certifier and preparer(s) for your facility. You should find the name of your facility below.

If you would like to view your facility's most current RMP, select the **View Current RMP** button.

A **First-time submission** requires that you enter information for all nine sections of the RMP. A **Section Selection** function will appear on the next page.

If you have previously submitted an RMP, your facility information will be pre-populated. In order to make a **Correction** to your existing RMP or a **Resubmission**, please click on the appropriate button for further action.

A **Correction** should be used to report administrative or other changes at your facility (e.g. changes in emergency contact information, facility address, or change in accident history). This does not require an update of your entire RMP. Submitting a corrected RMP does not change your five-year anniversary date. If your facility has not submitted an RMP Resubmission by its anniversary date, you will not be able to make an RMP Correction. You must submit an RMP Resubmission in this instance. The Correction button enables you enter the data for your facility and update the RMP as a correction.

A **Resubmission** is an update of all nine sections of your RMP. Revising your previous RMP appropriately and submitting the updated RMP Resubmission will re-set your five-year anniversary date. The Resubmission button enables you enter the data for your facility and update the RMP as a Resubmission.

WARNING! You can not use this system if you are reporting Confidential Business Information (CBI). Please contact the RMP Reporting Center at 301-429-5018 for questions regarding CBI reporting.

Selected Facility:

Facility ID	Facility Name	Anniversary Date	Transaction Type	Status
100000129381	Camie Campbell Ltd.	01/23/2014		

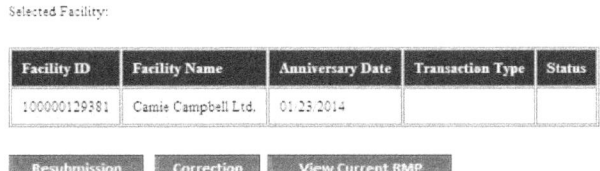

Select the appropriate button ("Correction" or "Resubmission") and you will have access to the RMP*eSubmit program to update your RMP.

CHAPTER 2 ENTERING DATA
(FIRST-TIME, CORRECTIONS, RESUBMISSIONS)

Section 1. Registration

All covered facilities must complete the registration portion of the RMP, even if they have a Program 1 process. The registration section requires facility identification information. You cannot enter data for Sections 2 through 5, 7, and 8 until you have entered process-specific information.

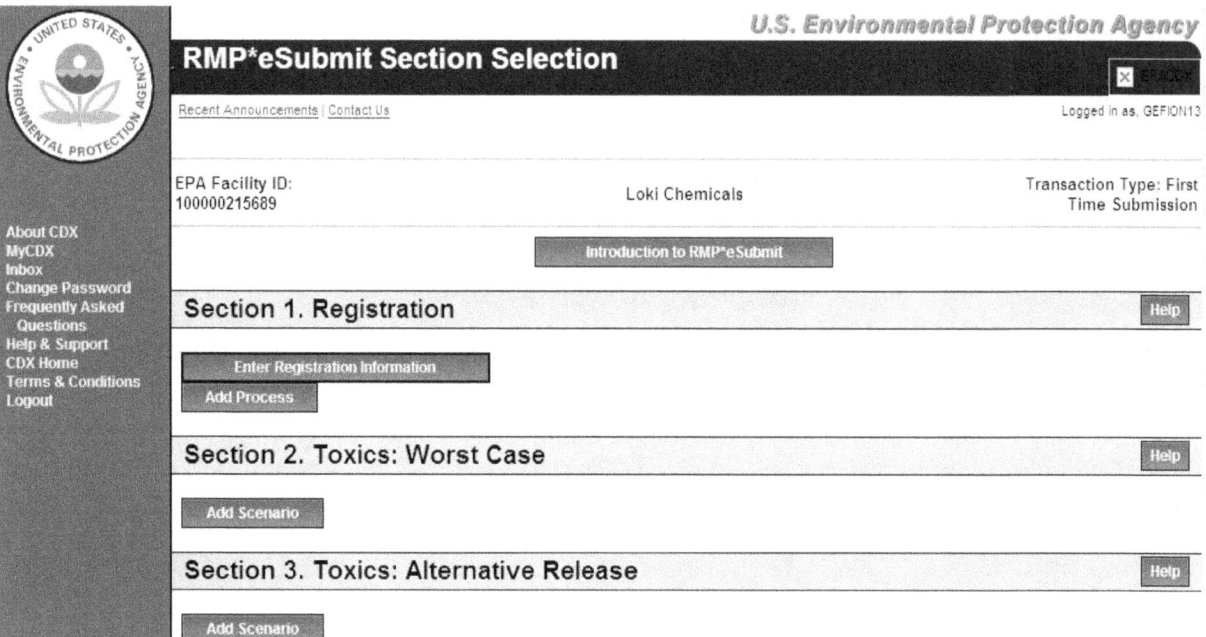

The following is a discussion of each element in Section 1. Registration. Click the "Enter Registration Information" button.

>   NOTE: Please remember to periodically use the "Save and Return" button at the bottom of the page. As a security measure, if you remain inactive in the system for 19 minutes, the system will log you out at the 20th minute and your entered data will not be saved! Each time you save, the system will return you to the RMP*eSubmit Section Selection page. Click the "Enter Registration Information" button to enter more data.

>   NOTE: If you do not wish to keep the information you've entered, click the "Discard Changes" button.

U.S. Environmental Protection Agency

RMP*eSubmit Section 1. Registration Information

Recent Announcements | Contact Us   Logged in as, GEFION13

| EPA Facility ID:
100000215689 | Loki Chemicals | Transaction Type: First
Time Submission |

Help

1.1. Source Identification:

1 1 a Facility name: *   Loki Chemicals

1 1 b Parent company #1 name: Norse Chemical Manufacturers

1 1 c Parent company #2 name:

1.2. EPA facility identifier: 100000215689

1.3. Other EPA systems facility ID:

1.4. Dun and BradStreet numbers (DUNS):

1 4 a Facility DUNS

1 4 b Parent company #1 DUNS

1 4 c Parent company #2 DUNS

1.5. Facility location address:

1 5 a Street 1 *   1234 Warehouse Way

About CDX
MyCDX
Inbox
Change Password
Frequently Asked
  Questions
Help & Support
CDX Home
Terms & Conditions
Logout

1.1   Source identification

1.1a.   Facility name:

Provide the name of your facility. The name must be specific to the site; if the site is part of a large corporation, the name may be the corporate name plus the location (for example, "ABC Chemicals - Hightown Plant").

Definition:

Throughout the Risk Management Program, the term facility means "any buildings, structures, equipment, installations or substance emitting stationary activities (i) which belong to the same industrial group, (ii) which are located on one or more contiguous properties, (iii) which are under the control of the same person (or persons under common control), and (iv) from which an accidental release may occur."

1.1 b.   Parent company #1 name:

The parent company is the corporation or other business entity that owns at least 50 percent of the voting stock of your company. If your facility is owned by a joint venture, enter the first two major owners here. If your facility does not have a parent company, leave this data element blank.

1.1c.   Parent company #2 name:

If your facility is owned by a joint venture, enter the name of the second major owner here.

1.2 EPA facility identifier (display only):

The facility ID number for your facility is displayed. *This is a pre-populated field that can't be edited.* If you need a new facility ID number, contact the RMP Reporting Center at 703-227-7650.

1.3 Other EPA systems facility identifier (only 15-characters allowed):

In EPA's efforts to streamline facility reporting, incoming facility information needs to be linked with existing facility information across different environmental programs. To help ensure your facility's RMP data is properly linked and that your facility does not receive multiple EPA facility IDs, please enter one of the following facility identification numbers in the following priority order:

- First, if your facility has a 15-characters Toxic Release Inventory identification number (**TRI ID**), enter it here. If your facility has a TRI ID, but you do not know the number, you can find it by searching the TRI database, http://www.epa.gov/enviro/facts/tri/customized.html or by calling the *EPA's Superfund, TRI, EPCRA, RMP & Oil Information Center* (also known as the *Info Center*), 1-800-424-9346.
- Second, if your facility does not have a TRI ID, but reports to EPA under another program and has a number referred to as a **FINDS** or Unique Identification Number (**UIN** with 12 characters), enter it here. If your facility has a FINDS or UIN number, but do not know what it is, you can find it by searching EPA's Facility Identification Initiative (FII) database, http://www.epa.gov/enviro/html/fii/fii_query_java.html.
- Third, use any one of the following numbers:
  - If your facility is covered by hazardous waste regulations under the Resource Conservation and Recovery Act (RCRA), enter your **RCRIS Handler ID** (12 characters). You can find your RCRIS ID via http://www.epa.gov/enviro/html/rcris/rcris_query_java.html (clicking this link opens a new browser window)
  - If your facility is covered by regulations under the Comprehensive Environmental Response, Compensation, and Liability Act (CERCLA), enter your **CERCLIS Site ID** (7 characters). You can find your CERCLIS ID via http://www.epa.gov/enviro/html/cerclis/cerclis_query.html
  - If your facility does not have an ID number as described above, leave this data element blank.

1.4 Dun and Bradstreet Numbers (DUNS):

1.4a. Facility DUNS:

The Data Universal Numbering System (DUNS) is a nine-digit identification number that allows your facility to be cross-referenced to various business information. If your facility has a DUNS number, it should be available from your treasurer or financial officer. You can also obtain the number from your local Dun and Bradstreet office (check the telephone book White Pages). If you do not have a DUNS number, leave this field blank.

1.4b.   Parent company #1 DUNS:

Provide DUNS number of your parent company(-ies), if applicable. If your facility is owned by a joint venture, provide the numbers for your two major owners. If your facility does not have a parent company or the parent company does not have a DUNS number, leave this field blank.

1.4c.   Parent company #2 DUNS:

Provide DUNS number of your second parent company, if your facility is owned by a joint venture. Otherwise, leave this field blank.

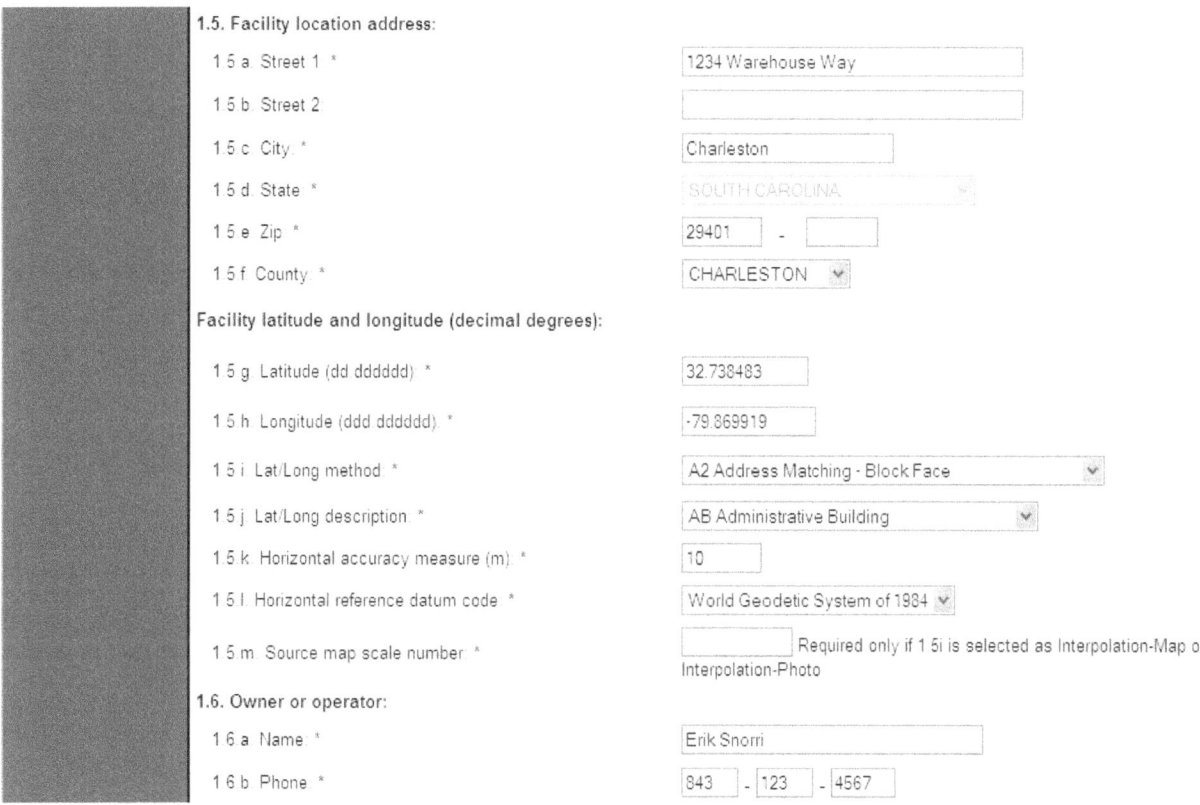

1.5   Facility location address:

Provide your facility *location* address, including the street, city, state, and ZIP code (including the 4 digit extension, if applicable. This is the location where regulated substances are present. The city should be the local legal jurisdiction, such as a township or village. Use local street and road designations, not post office or rural box numbers. **This location may not represent the mailing address. EPA will not attempt to mail correspondence to this address. All correspondence will be mailed to the address provided at 1.6 c-g.** If your facility location and mailing address are the same, then you will fill in the same address at 1.6 c-g.

1.5a. Street 1:

Provide your facility location address.

1.5b. Street 2:

Provide any additional facility location address information.

1.5c. City:

Provide the city name from your facility location address.

1.5d. State:

This is a pre-populated field and can't be edited.

1.5e. ZIP code:

Provide the ZIP code from your facility location address, including the 4-digit extension (if applicable).

1.5f. County:

The drop down list will populate with counties based upon the facility location state. Select a county from the drop down list that appears when you click the down arrow. If you edit this element with a county different from the county entered previously, you will have to select or enter a new LEPC in field "1.10 LEPC."

Facility latitude and longitude (decimal degrees):

1.5g. Latitude (dd.dddddd):

Provide the latitudinal coordinates of your facility. Facility latitude must be a value between -90 and 90. Decimal degrees use the format: +/- dd.dddddd where:

> "-" = west of the prime meridian (longitude) or south of the equator (latitude)

> "+" = east of the prime meridian (longitude) or north of the equator (latitude)

> NOTE: If you edit your facility's latitude, it must fall within the specific lat/long (latitude/longitude) boundary for your facility's address, or it will be rejected by RMP*eSubmit.

1.5h. Longitude (dd.dddddd):

Provide your facility longitude in decimal degrees. You must provide the longitudinal coordinates of your facility. Facility longitude must be a value between - 180 and 180. Decimal degrees use the format: +/- ddd.dddddd where:

> "-" = west of the prime meridian (longitude) or south of the equator (latitude)

> "+" = east of the prime meridian (longitude) or north of the equator (latitude)

> NOTE: If you edit your facility's longitude, it must fall within the specific lat/long boundary for your facility's address, or it will be rejected by RMP*eSubmit.

1.5i.   Lat/Long method:

You also must indicate the method that you used to determine your facility's latitude and longitude data. Select the method for which you obtained your lat/long coordinates from the drop down list.

1.5j.   Lat/Long description:

The table below lists the codes to be used for this element and provides a brief description of each method. The most common methods for determining latitude and longitude are I1 (Interpolation-Map), and I4 (Interpolation-Digital Map Source). Use I1 if you obtained your facility's latitude and longitude from a paper map. Use I4 if you obtained your facility's latitude and longitude from a computer-based geographic information system (GIS).

Code	Abbreviation	Description
AB	Administrative Building	A building, structure, or portion thereof that houses the administrative functions of a facility as opposed to production or manufacturing activities
AE	Atmospheric Emissions Treatment Unit	Equipment installed for the express purpose of treating chemical emissions prior to their release into the atmosphere
AM	Air Monitoring Station	Equipment installed at a predetermined location for the automatic, manual, or periodic collection of environmental air samples
AS	Air Release Stack	A free-standing vertical structure constructed for the conveyance and release of chemical emissions into the air
AV	Air Release Vent	A horizontal structure constructed for the release of chemical emissions into the air, typically from the side or roof of a building
CE	Center of Facility	A representative center point within the boundary of a facility
FC	Facility Centroid	The calculated center of a contiguous facility
IP	Intake Pipe	A pipe or intake opening constructed for the collection and conveyance of water
LC	Loading Area Centroid	The calculated center of a portion of a facility associated with loading activities
LF	Loading Facility	The portion of a facility associated with loading and/or transshipment activities
LW	Liquid Waste Treatment Unit	Equipment installed for the express purpose of treating chemical emissions prior to their release to water, publicly owned treatment works (POTW) or off-site transfer
NE	NE Corner of Land Parcel	The northeast most corner or boundary of a land parcel

Code	Abbreviation	Description
NW	NW Corner of Land Parcel	The northwest most corner or boundary of a land parcel
OT	Other	See descriptive comment field
PC	Process Unit Area Centroid	The calculated center of a portion of a facility associated with processing and/or manufacturing activities
PF	Plant Entrance (Freight)	The entrance to a facility associated with transshipment activities
PG	Plant Entrance (General)	The front gate or general entrance of a facility
PP	Plant Entrance (Personnel)	The entrance to a facility associated with employees
PU	Process Unit	The portion of a facility associated with processing and/or manufacturing activities
SD	Solid Waste Treatment/Disposal Unit	The portion of a facility associated with the treatment and/or disposal of solid waste
SE	SE Corner of Land Parcel	The southeast most corner or boundary of a land parcel
SP	Lagoon or Settling Pond	The portion of a facility designed to accommodate sedimentation or settling of chemical by-products necessitated by the manufacture, production, or use of chemicals
SS	Solid Waste Storage Area	The portion of a facility associated with the storage of solid waste
ST	Storage Tank	A receptacle or chamber used for storing bulk fuels or chemicals
SW	SW Corner of Land Parcel	The southwest most corner or boundary of a land parcel
UN	Unknown	
WA	Wellhead Protection Area	An area at the earth's surface buffering a wellhead
WL	Well	A shaft drilled in the earth for purposes such as obtaining subsurface drinking water, or collecting groundwater monitoring samples
WM	Water Monitoring Station	A location or study area for the automatic, manual, or periodic collection of water samples
WR	Pipe Release to Water	The point at which a pipe constructed for the conveyance and release of water-borne chemical emissions reaches a water body

1.5k. *Horizontal accuracy measure (in meters):*

This data provides a single, uniform statistical methodology for estimating the positional accuracy of points on maps and in digital spatial data. If you have difficulty obtaining this information, please contact the *EPA's Superfund TRI, EPCRA, RMP & Oil Information Center* at 1-800-424-9346 for additional guidance.

1.5l.   Horizontal reference datum code:

You must select the code from the available drop down list that represents the reference datum used in determining the lat/long coordinates you entered in 1.5 g and 1.5 h, respectively. If you have difficulty obtaining this information, please contact the EPA's Superfund *TRI, EPCRA, RMP & Oil Information Center* at 1-800-424-9346 for additional guidance.

The range of permissible values includes:

Horizontal Datum Code	Horizontal Datum Description
001	North American Datum of 1927
002	North American Datum of 1983
003	World Geodetic System of 1984

North American Datum Code of 1927

The North American Datum (NAD) of 1927 uses the Clarke 1866 spheroid to represent the shape of the earth. The origin of this datum is a point on the earth referred to as Meades Ranch in Kansas. Many NAD 1927 control points were calculated from observations taken in the 1800s. These calculations were done manually and in sections over many years. Therefore, errors varied from station to station.

North American Datum Code of 1983

The North American Datum of 1983 is based upon both earth and satellite observations, using the GRS80 spheroid. The origin for this datum is the earth's center of mass. This affects the surface location of all longitude–latitude values enough to cause locations of previous control points in North America to shift, sometimes as much as 500 feet. A 10-year multinational effort tied together a network of control points for the United States, Canada, Mexico, Greenland, Central America, and the Caribbean.

Because NAD 1983 is an earth-centered coordinate system, it is compatible with global positioning system (GPS) data. The raw GPS data is actually reported in the World Geodetic System 1984 (WGS 1984) coordinate system.

World Geodetic System of 1984 (WGS84)

The World Geodetic System of 1984 is the reference frame used by the U.S. Department of Defense and is defined by the National Imagery and Mapping Agency (formerly the Defense Mapping Agency). WGS 84 is used by Department of Defense for all its mapping, charting, surveying, and navigation needs, including its GPS "broadcast" and "precise" orbits. The latest revision of WGS 84 was in 2004.

1.5m. Source map scale number:

This is the proportional distance on the ground for one unit of measure on a map or photo. This information must be supplied if you have chosen a lat/long method of type **Interpolation - Map** (I1) or **Interpolation - Photo** (I2). If you have difficulty obtaining this information, please contact the EPA's Superfund *TRI, EPCRA, RMP & Oil Information Center* at 800-424-9346 for additional guidance.

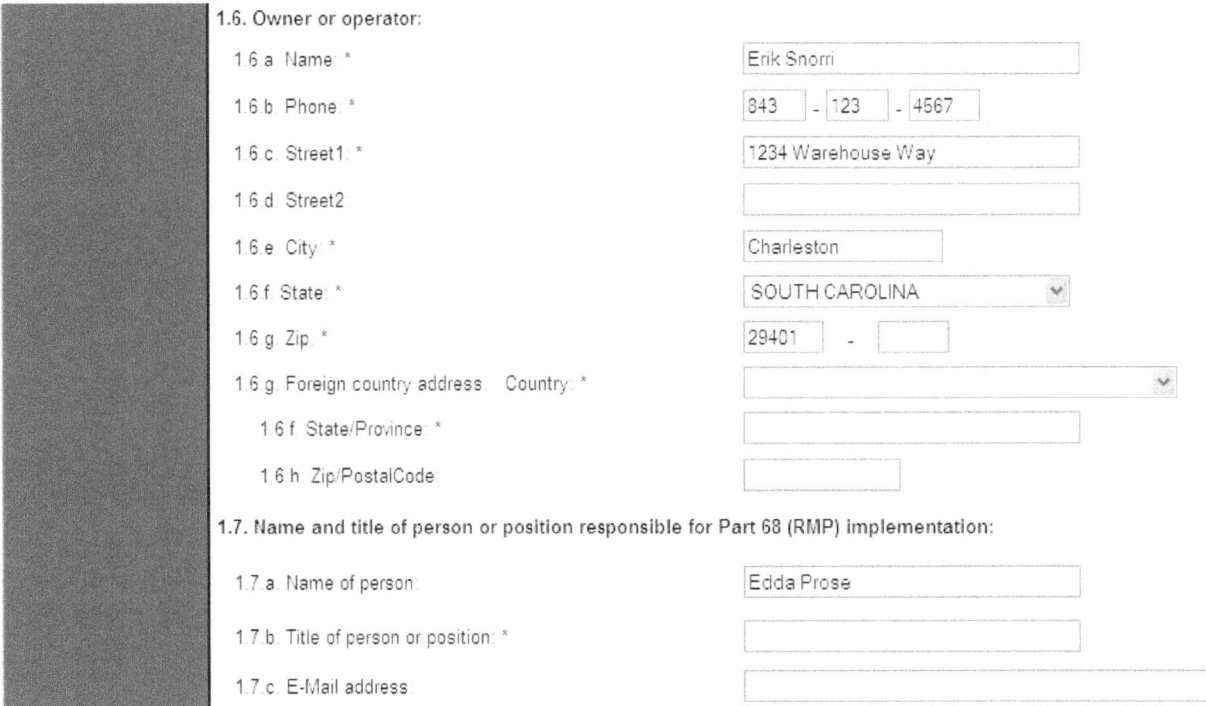

1.6   Owner or operator:

1.6a.   Name:

This is the name of the legal owner or operator of the facility (person, company, association, or government agency).

1.6b.   Phone:

Provide the facility owner or operator's business phone number, including area code.

1.6c.   Street 1:

Provide the facility owner or operator's business mailing address, street – line 1. In this instance, you may use post office and rural box numbers, if appropriate. **This is the address to which all correspondence will be mailed.**

1.6d.   Street 2:

Provide the facility owner or operator's business mailing address, street – line 2. In this instance, you may use post office and rural box numbers, if appropriate. **This is the address to which all correspondence will be mailed.**

1.6e.   City:

Provide the facility owner or operator's business mailing address, city. The city should be the local legal community, for example, a township or village. **This is the address to which all correspondence will be mailed.**

1.6f.   State:

Use the drop down list to select the state of the facility owner or operator's business mailing address. **This is the address to which all correspondence will be mailed.** You may also select a state from the list that appears when you click the down arrow.

1.6g.   ZIP code:

Provide the facility owner or operator's business mailing ZIP code (including 4-digit extension, if applicable). **This is the address to which all correspondence will be mailed.** If the address is in a foreign country, select the country, then provide the state/province and postal code.

1.6g.   Foreign country address (if applicable):

Country:

Provide the owner or operator's foreign business mailing address country.

1.6.f.   State/Province:

Provide the owner or operator's foreign business mailing state or province.

1.6.h.   ZIP/Postal code:

Provide the owner or operator's foreign business mailing address ZIP/postal code.

1.7   Name and title of person responsible for RMP 40 CFR Part 68 implementation:

1.7a.   Name of person:

Provide the name of the person with overall responsibility for the Risk Management Program at your facility site. Although the individual's name is not required, the title of the person or the position that has this responsibility is required.

1.7b.   Title of person or position:

Provide the title of the person or position with overall responsibility for the Risk Management Program at your facility site. Although the individual's name is not required, the title of the person or the position that has this responsibility is required.

1.7c.   Email address (optional):

Provide the email address of the person or position with overall responsibility for the Risk Management Program at your facility site. The individual's email address is not required.

1.8.a. Name: *	[ ]
1.8.b.Title: *	[ ]
1.8.c. Phone: *	[ ] - [ ] - [ ]
1.8.d. 24-hour phone: *	[ ] - [ ] - [ ]
1.8.e. Ext. or PIN:	[ ]
1.8.f. E-Mail address: *	[ ] (enter 'N/A' if not applicable)
1.9. Other points of contact:	
1.9.a. Facility or parent company E-Mail address:	[ ]
1.9.b. Facility public contact phone:	[ ] - [ ] - [ ]
1.9.c. Facility or parent company www homepage address:	[ ]
1.10. LEPC:	Other/Unknown ▾
Specify if Other selected:	[ ]
1.11. Number of fulltime equivalent employees on site: * CBI	[_____] or Claim as CBI? ☐
1.12. Covered by:	
1.12.a. OSHA PSM:	☐
1.12.b. EPCRA section 302:	☐
1.12.c. CAA Title V:	☐

1.8   Emergency contact:

1.8a.   Name:

Provide the name of the person designated as the emergency contact. If your facility has a regulated toxic substance, you may already have designated a facility emergency coordinator in your notification to your Local Emergency Planning Committee (LEPC) under EPCRA Section 312. If your facility has more than one emergency contact, provide just one name for this entry. You may want to list the other emergency contacts in your Executive Summary.

Your emergency contact name should be:

- An employee or contract employee of your facility;
- Knowledgeable about your site;
- Aware of all emergency plans and procedures; and
- Able to provide emergency response support or be able to direct/assist emergency response personnel to provide support.

1.8b.   Title:

Provide the title or job classification of the person designated as the emergency contact. If you have a regulated toxic substance, you may already have designated a facility emergency coordinator in your notification to your LEPC under EPCRA Section 312.

1.8c.   Phone:

Provide the phone number, including area code, where the emergency contact can be reached during normal business hours. You will most likely provide the facility telephone number here. If your facility does not have a phone number, you may provide the business phone number of the emergency contact, the phone number of a dispatcher, or the customer service phone number.

1.8d.   24-hour phone:

Provide a 24-hour phone number for your facility.

1.8e.   Ext. or PIN:

Provide an extension or PIN #, in this space, if applicable.

1.8f.   Email address:

Provide the email address for the emergency contact who can be reached during normal business hours. If your facility does not have an emergency contact email address, enter N/A.

1.9   Other points of contact (optional):

1.9a.   Facility or parent company email address (optional):

You may provide an email address to which inquiries from the public could be sent. The email address could be for the person who developed your RMP or your public liaison office.

1.9b.   Facility public contact phone (optional):

You may provide a phone number for public inquiries. It could be the phone number of the person who developed your RMP or of your public liaison office.

1.9c.   Facility or parent company (www. homepage address (optional):

You may provide the Internet address where you have details on your accident prevention program or other facility/corporate information.

1.10   LEPC:

Select the name of your LEPC for your planning district from the drop down list. LEPCs were created to do local planning under the Emergency Planning and Community Right to Know Act (EPCRA) of 1986. In RMP*eSubmit, you will select your LEPC's name from a drop down list based on the facility county information that you have entered. If you are unsure or do not know the LEPC associated with your county, select Other/Unknown.

Specify if Other selected:

If the LEPC that represents your county does not appear on the drop down list and you know the name of your LEPC, select Other/Unknown and type the LEPC name.

> **NOTE:** *If you do not know your LEPC's name, you can contact your local fire department or your State Emergency Response Commission (SERC):*
> *http://www2.epa.gov/epcra/state-emergency-response-commissions*

1.11   Number of full-time equivalent employees on site:

Provide the number of full-time equivalent employees who work at your facility. To determine the number of full-time equivalent employees at your facility, add together the fractions of full-time work performed by part-time or seasonal employees and round to the nearest whole number. Do not include contract employees. If your facility is unmanned or is only staffed by part-time employees, you should briefly explain these circumstances in the Executive Summary.

1.12. Covered by:

1.12.a. OSHA PSM: ☐

1.12.b. EPCRA section 302 ☐

1.12.c. CAA Title V. ☐

1.12.d. Air operating permit ID [＿＿＿＿]

1.13. OSHA Star or Merit Ranking: ☐

1.14. Last safety inspection (by an external agency) date: [＿＿＿] [▦]

1.15. Last safety inspection performed by an external agency: *   [Other ▾]

Specify if Other selected [＿＿＿＿]

1.16. Will this RMP involve Predictive Filing?: ☐

1.17. Processes   To register covered processes return to Section Selection page, Section 1: Add Process button

1.18. RMP Preparer Information:

If an outside contractor prepared this risk management plan, please enter information concerning this contractor in the fields below. If you enter an outside contractor name in 1.18.a then you must complete 1.18.b through 1.18.f.

1.18.a. Name [＿＿＿＿]

1.18.b. Telephone: * [＿] - [＿] - [＿]

1.18.c. Street1: * [＿＿＿＿]

1.18.d. Street2 [＿＿＿＿]

1.18.e. City: * [＿＿＿＿]

1.12   Covered by:

1.12a. OSHA's PSM:

This element refers to OSHA's Process Safety Management (PSM) of Highly Hazardous Chemicals Standard (29 CFR 1910.110). This data element applies to your facility as a whole and is not a process-by-process determination. Therefore, if any process at your facility is subject to the OSHA PSM standard, check this box, even if the PSM process is not covered by the RMP rule.

> **NOTE:** *For further information about OSHA's PSM standard, visit:*
> *http://www.osha.gov/SLTC/processsafetymanagement/*

1.12b. EPCRA Section 302:

If you have more than a threshold quantity of a substance that is an Extremely Hazardous Substance (EHS) on site, your facility is subject to EPCRA Section 302 notification requirements (a one-time notification to the State Emergency Response Commission (SERC) and LEPC that states your facility has one of the listed EHS on site). If your facility is subject to those requirements, check this box, regardless of whether the EHS is covered by the RMP rule or is held in a process that is below the Section 112(r) threshold quantity. Two quick hints:

- If your facility is subject to the RMP rule because you have more than a threshold quantity of a toxic substance listed under Section 112(r), you are subject to EPCRA Section 302. Select this data element.
- If your facility is subject to the RMP rule only as a result of flammable substances, you are not subject to EPCRA Section 302. Leave this data element blank.

1.12c. CAA Title V:

If your facility has a Title V operating permit, check this box.

1.12d. Air operating permit ID:

If your facility has an air operating permit ID, provide the ID number.

1.13 OSHA star or merit ranking (optional):

Check this box if your facility has received a star or merit ranking under OSHA's Voluntary Protection Program. Your facility is exempt from audits conducted pursuant to Section 68.220(b)(2) and (7).

1.14 Last safety inspection (by an external agency) date:

Provide the date of the facility's last safety inspection by an external agency.

1.15 Last safety inspection performed by an external agency:

Select the agency (or agencies) that performed the inspection from the drop down list. Choose from the following:

- EPA
- Fire department
- Never had one
- OSHA
- State environmental agency
- State occupational safety organization

  Specify if Other selected:

  Provide the external agency if not present in element 1.15. If the last safety inspection was a joint inspection, enter multiple agencies. If the agency does not appear in the drop down list in element 1.15, select the Other option and provide the name of the agency.

1.16   Will this RMP involve Predictive Filing?

Predictive Filing is an RMP filing option that allows your facility to submit an RMP which includes regulated substances which may not actually be present at the facility at the time the RMP is submitted. This option is intended to assist facilities (such as chemical warehouses, chemical distributors, batch processors, and the like) whose operations involve highly variable types and quantities of regulated substances, but who are able to forecast their inventory with some degree of accuracy. Under 40 CFR Section 68.190, a facility is required to update and resubmit its RMP no later than the date on which a new regulated substance is first present in a covered process above a threshold quantity. By using Predictive Filing, you will not be required to update and resubmit your RMP when you receive a new regulated substance if that substance was included in your latest RMP submission (as long as you receive it in a quantity that does not trigger a revised offsite consequence analysis as provided in 40 CFR Section 68.36).

If you use Predictive Filing, you should implement your Risk Management Program and prepare your RMP in exactly the same way as if all of the substances included in the RMP were actually present. This means that you must meet all rule requirements for each regulated substance for which you file, whether or not that substance is actually present onsite at the time you submit your RMP. Depending on the substances for which you file, this may require you to perform additional worst-case and alternative-case scenarios and to implement additional prevention program elements.

> NOTE : If your facility uses this option, you must still update and resubmit your RMP if you receive a new regulated substance which was not included in your latest RMP. Your facility must also continue to comply with the other update requirements stated in 40 CFR 68.190.

If your facility uses Predictive Filing, the RMP database for your facility will indicate that your facility has filed a predictive RMP. This will indicate that some of the chemicals in your RMP may not actually be present onsite, but will not indicate which specific chemicals are onsite at any given time. Therefore, you may receive more frequent questions from the public, local officials, or implementing agencies about your actual chemical inventory. EPA encourages you to engage in more frequent dialogue with these parties, and in particular with local emergency planners, emergency responders, and community officials to update them on your current inventory of regulated substances. Check this box if your RMP involves Predictive Filing. This information can't be edited for an RMP Correction.

1.17   Processes

The following information includes a description of all elements in Section 1.17 Registration: Processes.

*NOTE: Information on **Processes** will be entered on the Add Process page. Click the "Add Process" button under "Section 1. Registration" on the RMP*eSubmit Section Selection page. **WARNING!** If you delete a chemical or an entire process record in Section 1, processes, all chemical records linked to the deleted chemical or deleted process in sections 2, 3, 4, 5, 7 and 8 will also be deleted. Program level options cannot be selected or edited in an RMP Correction.*

Process ID: The process ID is a system generated number which has no bearing on your submission.

Process description (optional):

This description helps you to track multiple processes as you fill out the remaining sections of your RMP. Provide your process description in this element.

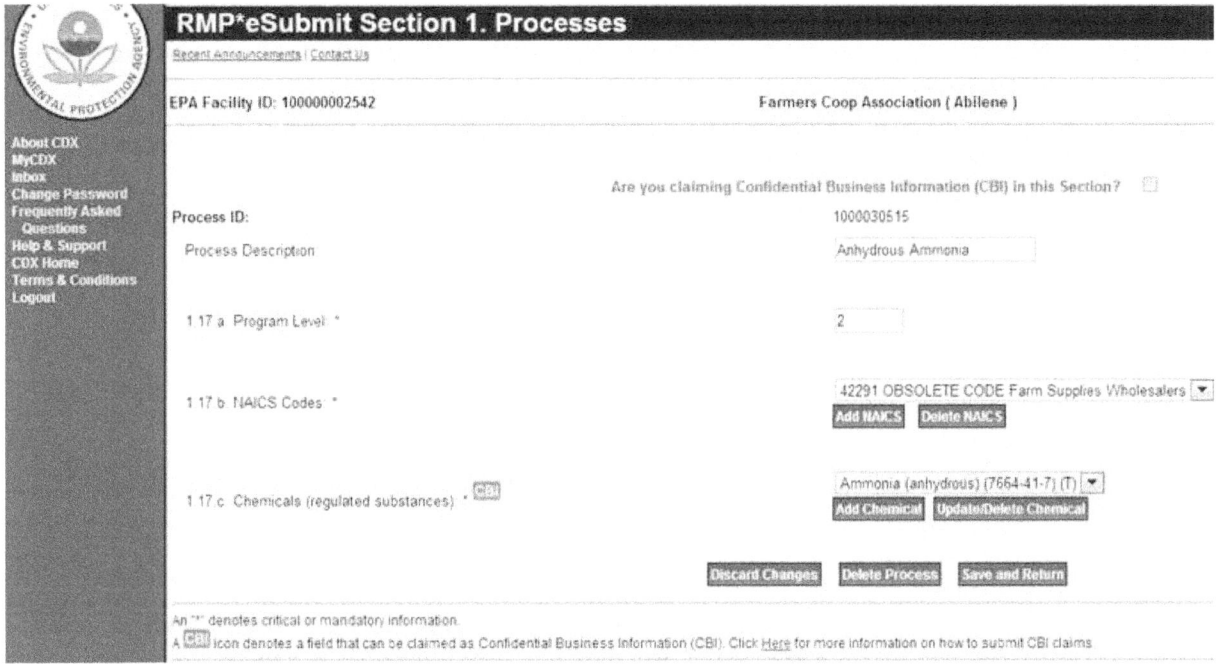

1.17a. Program level:

Enter or edit the program level that corresponds to each process. The rule imposes different requirements on processes based on the potential for public impacts and the level of effort needed to prevent accidents. EPA has set three levels of requirements that apply to covered processes:

Program level 1	Program level 2	Program level 3 (unless eligible for Program 1)
Process has experienced no accident in past 5 years that resulted in significant off-site impacts, No public receptors in worst-case circle, and Emergency response coordinated with local responders.	Process is not eligible for Program 1 or subject to Program 3.	Process is subject to OSHA PSM, or Process is in NAICS code 32211, 32411, 32511, 32518, 32518, 32519, 325199, 325211, 325311, or 32532.

WARNING! Changing the Program Level will delete other sections in the RMP that depend on its value. For example, changing the Program Level from 2 to either 1 or 3 will delete the Section 8 Program corresponding to the Program Level 2 Process, if one exists. Changing the Program Level from 3 to either 1 or 2 will delete the Section 7 Program corresponding to the Program Level 3 Process, if one exists.

If you can qualify a process for Program 1, it is in your best interests to do so, even if the process is already subject to OSHA PSM. For Program 1 processes, the implementing agency will inspect and enforce only on compliance with the minimal Program 1 requirements. If you assign a process to Program 2 or 3 when it might qualify for Program 1, the implementing agency will inspect or enforce for compliance with all the requirements of the higher program levels. If, however, you are already in compliance with the prevention elements of Program 2 or Program 3, you may want to use the RMP to inform the community of your prevention efforts. *The Program Level cannot be edited for an RMP Correction.*

KEY POINTS TO REMEMBER

In determining program levels for your process(es), keep in mind the following:

1. **The program levels apply to individual processes** and generally indicate the risk management measures necessary to comply with this regulation for the process, not the facility as a whole. The eligibility of one process for a program level does not influence the eligibility of other covered processes for other program levels.
2. **Any process can be eligible for Program 1,** even if it is subject to OSHA PSM or is in one of the NAICS (North American Industry Clarification System) codes subject to Program 3.
3. **Program 2 is the default program level.** There are no "standard criteria" for Program 2. Any process that does not meet the eligibility criteria for either Programs 1 or 3 is subject to the requirements for Program 2.

Refer to Chapter 2 of the *General Guidance for Risk Management Programs* for more information on determining the Program levels of your processes. Once you determine the program level, simply enter 1, 2, or 3 for this data element.

1.17b. NAICS codes:

If you know your facility's NAICS code, enter your NAICS code. To add a NAICS code to your process, click the "Add NAICS" button, which will take you to the Select NAICS to add screen. To delete the existing NAICS code, click the "Delete NAICS" button.

1.17c. Chemicals (regulated substances):

Enter your chemical in this field. For each covered process, provide the names of all regulated substances held above the threshold. Many regulated substances have synonyms. However, you must enter the name of the regulated substance as it appears in Section 68.130. If you have a NFPA-4 flammable mixture containing regulated flammables, you may list it as a "flammable mixture." List all of the regulated substances contained in the mixture; however, only report the quantity of the entire mixture, not the individual substances. RMP*eSubmit contains a pick list of all regulated substances. *For an RMP Correction, a chemical cannot be added; only the Chemical Quantity (in pounds) can be updated.*

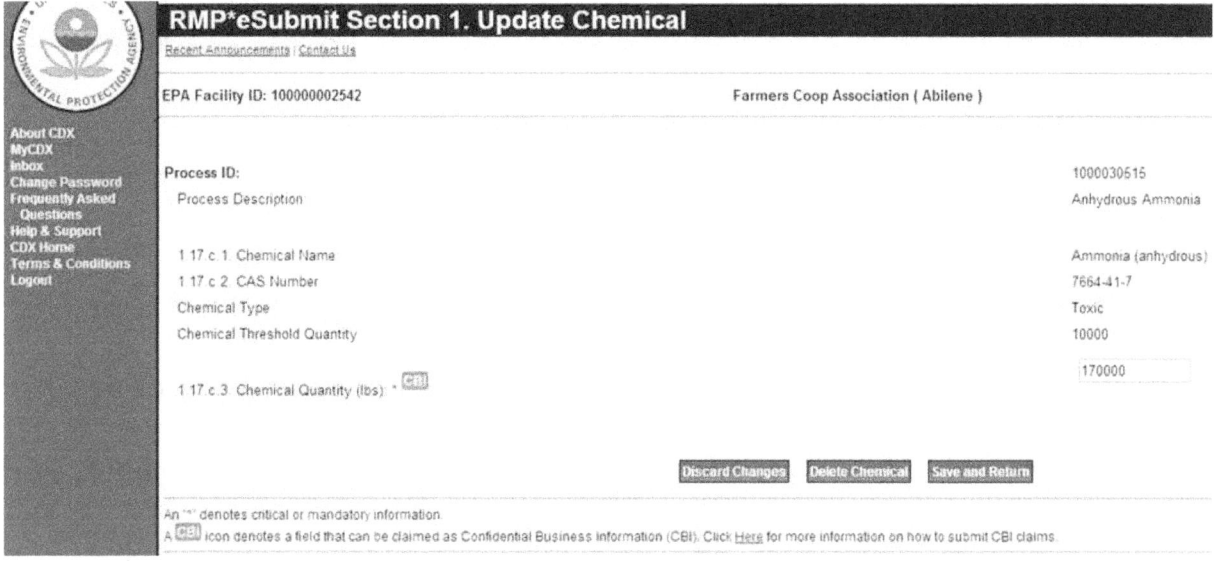

Chemical Update

1.17c.1 Chemical name:

The chemical name field cannot be edited.

1.17c.2 CAS number:

The CAS number field cannot be edited.

Chemical type:

The chemical type field cannot be edited.

Chemical threshold quantity:

The chemical threshold quantity cannot be edited.

1.17c.3Chemical quantity (lbs):

For each chemical reported in 1.17 c.1, estimate the maximum quantity (in pounds) held in the covered process at any one time during the calendar year to two significant digits. For example:

>5,333 pounds should be reported as 5,300 pounds

>128,000 pounds should be reported as 130,000 pounds

NAICS Code(s)

The North American Industry Classification System (NAICS) categorizes businesses by fitting them into descriptive categories that correspond to five-digit or six-digit codes. NAICS codes replaced SIC codes in 1997. For this data element you should provide the NAICS code that most closely corresponds to the process; it will not necessarily be the same NAICS code for your facility as a whole. You may also enter additional NAICS codes if you wish to identify other aspects of a process not captured by the NAICS codes for the primary activity.

You should determine the NAICS codes for your processes based on your activities on site using the *2012 North American Industry Classification System Manual,* which can be viewed at http:www.census.gov/epcd/www/naics.html.

Once you have selected the sector that most appropriately reflects the sector for your facility (Step 1), you must select the subsector, which enables you to select a more specific sector within your industry (Step 2). Next, you must select the industry group that represents your facility (Step 3). The last option is to select the NAICS code that reflects your facility (Final Step).

1. The **Step 1 – Select Sector** drop down list enables you to add the sector of the NAICS code to Section 1. Processes.
2. The **Step 2 – Select Subsector** drop down list enables you to add the subsector of the NAICS code to Section 1. Processes.
3. The **Step 3 – Select Industry Group** drop down list enables you to add the industry group of the NAICS code to Section 1. Processes.
4. The **Final Step – Select NAICS code** drop down list enables you to select your NAICS code to Section 1. Processes.

Process ID:

The process ID is automatically generated by the RMP*eSubmit system and this information can't be edited.

Process name:

The process name can't be edited in this section. The process name can be edited in the Section 1. Processes screen.

Program level:

The program level can't be edited in this screen. The program level is set when you add or edit a process.

1.18. RMP Preparer Information:

If an outside contractor prepared this risk management plan, please enter information concerning this contractor in the fields below. If you enter an outside contractor name in 1.18.a then you must complete 1.18.b through 1.18.f.

1.18.a. Name

1.18.b. Telephone *

1.18.c. Street1 *

1.18.d. Street2

1.18.e. City *

1.18.f. State *

1.18.g. Zip *

1.18.g. If foreign country address   Country *

    1.18.f. State/Province *

    1.18.h. Zip/Postal Code

[Discard Changes] [Save and Return]

An "*" denotes critical or mandatory information.

A **CBI** icon denotes a field that can be claimed as Confidential Business Information (CBI). Click Here for more information on how to submit CBI claims.

1.18 RMP Preparer Information

1.18a. Name:

Provide the name of the person who prepared the Risk Management Plan for the facility. If you provide the name of the Preparer, you must provide the remaining contact information in fields 1.18 b-g.

1.18b. Telephone:

Provide the telephone number of the person who prepared the Risk Management Plan for the facility.

1.18c. Street 1:

Provide the street address of the person who prepared the Risk Management Plan for the facility.

1.18d. Street 2:

Provide any additional street address information for the person who prepared the Risk Management Plan for the facility.

1.18e. City:

Provide the city of the person who prepared the Risk Management Plan for the facility.

1.18f. State:

Select the state of the person who prepared the Risk Management Plan for the facility from the drop down list.

1.18g. ZIP code:

Provide the ZIP code of the person who prepared the Risk Management Plan for the facility.

1.18g.If Foreign country address: Country:

If the RMP Preparer is located in a foreign country, select the appropriate country from the drop down list.

1.18f. State/Province:

If the RMP Preparer is located in a foreign country, provide the appropriate country state/province.

1.18h. ZIP/Postal code:

If the RMP Preparer is located in a foreign country, provide the appropriate country ZIP/postal code.

Section 2. Toxics: Worst-Case

Worst-case release scenario analysis of covered processes as follows:

Report one worst-case release scenario for each Program 1 process. Program 1 processes must have no public receptors within the distance to the endpoint in the worst-case analysis.

If your facility has Program 2 or Program 3 processes, report <u>one</u> worst-case release scenario to represent ***all*** Program 2 and Program 3 processes having toxic regulated substances present above the threshold quantity, ***and*** <u>one</u> worst-case release scenario to represent ***all*** Program 2 and Program 3 processes having flammable regulated substances present above the threshold quantity. If you have more than one Program 2 or 3 process, you will report the worst-case release scenario for the Program 2 or 3 process that would have the greatest potential impact on the public (i.e., the greatest distance to endpoint). You may also need to submit an additional worst-case scenario for either hazard class (i.e., toxic or flammable), if a worst-case release from elsewhere at your facility would potentially affect a different set of public receptors than those affected by your initial worst-case scenario(s).

You may include one graphic (map or diagram) in electronic format for each release scenario that you report, but it is not required.

Complete this section for each toxics worst-case scenario you report.

The following is a discussion of each element in *Section 2. Toxics: Worst Case.*

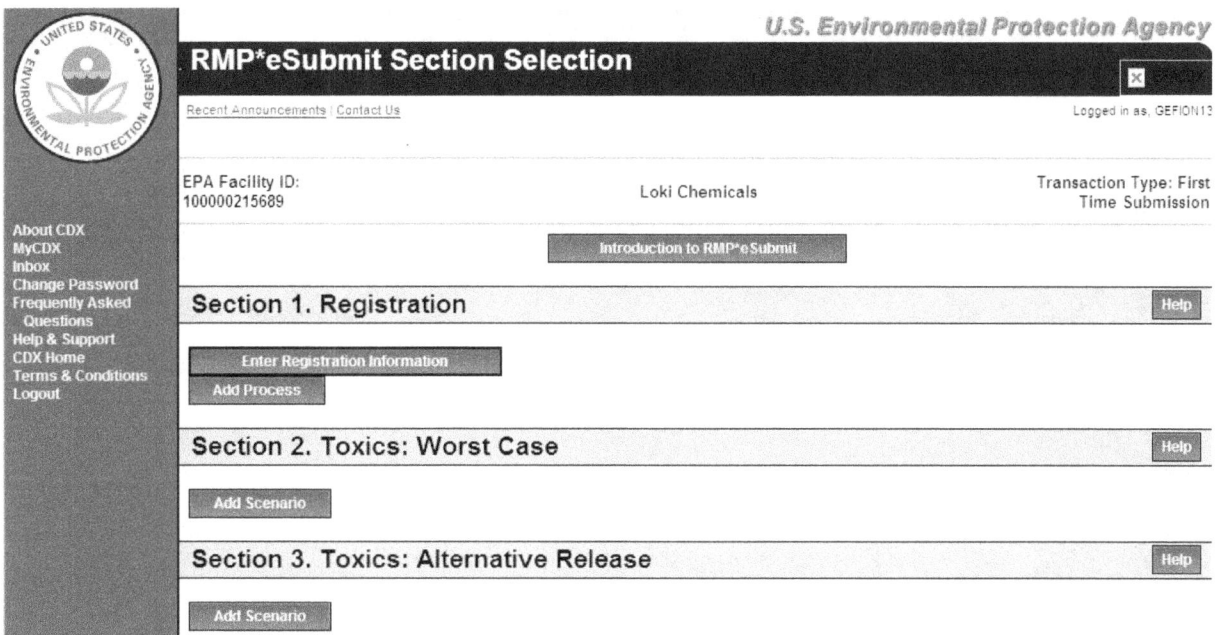

Select the "Add Scenario" button under *Section 2.Toxics: Worst Case*

> NOTE: Please remember to periodically use the "Save and Return" button at the bottom of the page. As a security measure, if you remain inactive in the system for 19 minutes, the system will log you out at the 20th minute and your entered data will not be saved! Each time you save, the system will return you to the RMP*eSubmit Section Selection page. Click the "Enter Registration Information" button to enter more data.

> NOTE: If you do not wish to keep the information you've entered, click the "Discard Changes" button.

Select the "Continue Add Scenario" button.

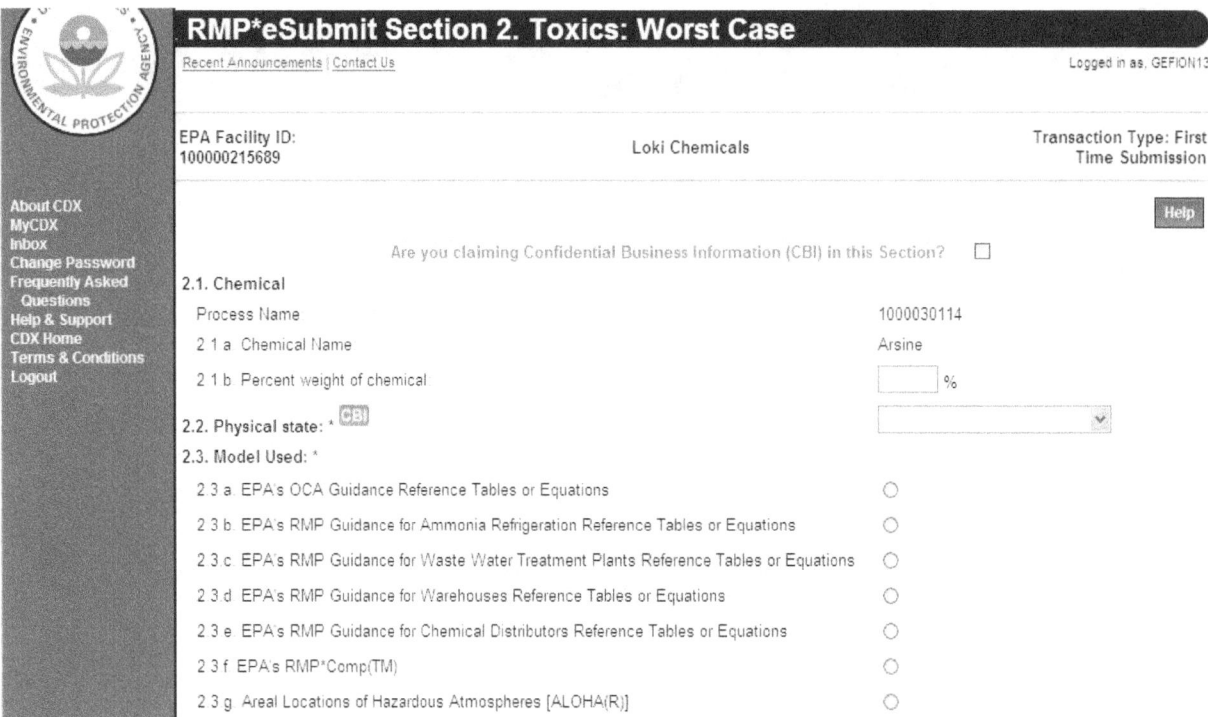

If your facility is claiming Confidential Business Information (CBI) in this section, check the box. If claiming CBI and you've checked the box, a pop-up message will appear describing the step that must be taken by the Preparer.

2.1   Chemical

a. Process name:

This field contains the process description from which the chemical was entered in Processes 1.17. This field is disabled and cannot be edited as it reflects information entered in *Section 1. Processes.*

b. Chemical name:

This field contains the regulated toxic chemical you evaluated in the worst-case scenario previously entered in Processes 1.17. This field is disabled and cannot be edited as it reflects information entered in *Section 1.17 Processes.*

c. Percentage weight of chemical:

If your worst-case scenario involves the release of a mixture containing a regulated substance, enter the percentage weight of the regulated substance in the mixture. (Leave blank if it is not a mixture.)

2.2   Physical state:

Select the physical state of the chemical as it is released in the scenario from the drop down list.

- Gas (Select if the chemical is a gas)
- Liquid (Select if the chemical is a liquid)
- Gas liquified by pressure
- Gas liquified by refrigeration

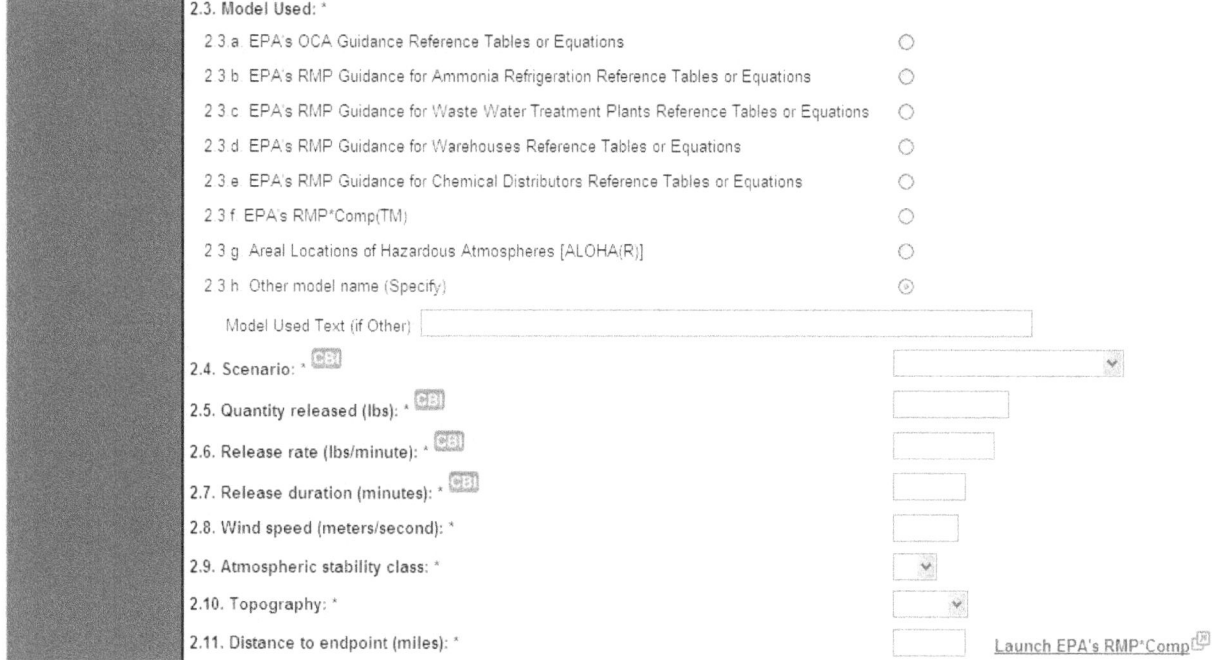

2.3   Model used (select one):

Select one of the options for the source of your results for your worst-case release. If you select the *Other model name (specify)* option, you must specify the other model name in the *Model Used Text (if Other)* field. The options are listed below:

a. EPA's *OCA Guidance* Reference Tables or Equations
b. EPA's *RMP Guidance for Ammonia Refrigeration Reference* Tables or Equations (now in *General Risk Management Program Guidance*)
c. EPA's *RMP Guidance for Waste Water Treatment Plants* Reference Tables or Equations (now in *General Risk Management Program Guidance*)
d. EPA's *RMP Guidance for Warehouses* Reference Tables or Equations
e. EPA's *RMP Guidance for Chemical Distributors* Reference Tables or Equations
f. EPA's RMP*Comp™
g. Areal Locations of Hazardous Atmospheres (ALOHA ®) (for Toxics only)
h. Other model name (specify)

2.4   Scenario:

Select one of the following that describes your worst-case release scenario from the drop down list:

- **Gas Release:** A gas release of the substance in a vapor state. If you hold a gas liquified under refrigeration, report the release as a liquid.

- **Liquid Spill and Vaporization:** A release of the substance in a liquid state with subsequent vaporization.

2.5   *Quantity released (lbs):*

Enter the quantity of toxic chemical you used for your worst-case scenario analysis in pounds in whole numbers. If you have less than one pound released, round up to one pound. You may want to clarify this in your *Executive Summary.*

2.6   *Release rate (lbs/minute):*

Enter the rate of release to the outside air in pounds per minute in whole numbers. For example:

     4.3 pounds should be reported as... 4 pounds

     19 pounds should be reported as... 20 pounds

See the General Risk Management Program Guidance for more information.

2.7   *Release duration (minutes):*

Enter the length of time in minutes for the entire quantity from the vessel, pipeline, or other source to be released to the outside air.

For gases, a gas liquified by pressurization alone, or a gas liquified by refrigeration where the released refrigerated liquid forms a pool of 1 cm or less in depth, you should assume that the release duration is 10 minutes.

For a liquid or a gas liquified by refrigeration where the released refrigerated liquid forms a pool deeper than 1 cm, the release duration should be the time required for a pool formed by the released substance to completely vaporize.

Although in some cases it may take longer than 60 minutes for the pool to completely volatilize, most dispersion models use the release rate and calculate the maximum downwind dispersion distance within 60 minutes. Therefore, you may enter 60 minutes for your duration, even if the duration from your modeling is longer than 60 minutes. You can also enter the exact duration from your modeling up to 9999.9 minutes.

2.8   *Wind speed (meters/second):*

If you used the OCA guidance or one of EPA's model program guidance documents, enter or edit 1.5 meters per second. If you modeled your scenario separately, enter or edit the wind speed used.

2.9   *Atmospheric stability class:*

If you used EPA's *OCA Guidance* Reference Tables or Equations or one of EPA's model program guidance documents, list select *F stability* from the drop down list. If you modeled your scenario separately, select the appropriate stability class used from the drop down list.

2.10 Topography:

Select whether the local topography is *urban* or *rural* from the drop down list. Urban means that there are many obstacles in the immediate area; obstacles include buildings or trees. Rural means that the terrain is generally flat and unobstructed in the immediate area.

2.11 Distance to endpoint (miles):

Enter the distance to the endpoint in miles to 2 significant digits, using the endpoint specified for the chemical in 40 CFR Part 68, Appendix A. Convert your modeling results into miles by dividing the distance in feet by 5280 or yards by 1760. Refer to the following to determine 2 significant digits:

> 0.397 miles should be reported as 0.40 miles
>
> 9.345 miles should be reported as 9.3 miles
>
> 20.764 miles should be reported as 21.0 miles

2.12 Residential population within distance to endpoint:

Enter the estimated population within the circle with a center at the point of the release and a radius determined by the distance to the endpoint to two significant digits (e.g., 5,500 people rather than 5,483). Population estimates include only residential populations.

2.13. Public receptors within distance to endpoint	
2.13.a. Schools	☐
2.13.b. Residences	☐
2.13.c. Hospitals	☐
2.13.d. Prison/Correctional facilities	☐
2.13.e. Recreational areas	☐
2.13.f. Major commercial, office or industrial areas	☐
2.13.g. Other	
2.14. Environmental receptors within distance to endpoint	
2.14.a. National or State parks, forests, monuments	☐
2.14.b. Officially designated wildlife sanctuaries, preserves or refuges	☐
2.14.c. Federal wilderness area	☐
2.14.d. Other	

2.13 Public receptors within distance to endpoint:

Select one or more of the public receptors within distance to endpoint by checking the box that corresponds with the specified receptor in fields 2.13 a-g.

a. **Schools:** Public and private elementary, secondary, or post-secondary educational institutions (e.g., colleges).
b. **Residence:** Public and private residences and dwellings wherever people live.

c. **Hospitals:** Public and Private Hospitals. Places that provide emergency care and or long term care of the sick or elderly.

d. **Prisons/Correction facilities:** Prisons where men or women are incarcerated. Holding places or state of confinement for criminals.

e. **Recreation areas:** Recreation areas include stadiums, parks, and public pools.

f. **Major commercial, office industrial areas:** Commercial, office, or industrial areas include industrial parks, office buildings, shopping malls, commercial areas, and commercial farms.

g. **Other:** Include other types of public receptors.

2.14   Environmental receptors within distance to endpoint:

Select one or more of the environmental receptors within distance to endpoint by clicking the check box that corresponds with the specified receptor in fields 2.14 a-d. These options are a combination of national or state parks, forests, or monuments which are within a circle whose center is the point of the release and the radius is determined by the distance to the endpoint. Select all that apply.

a. National or state parks, forests, or monuments

b. Officially designated wildlife sanctuaries, preserves, refuges

c. Federal wilderness areas

d. Other (Specify). (Include other types of environmental receptors.)

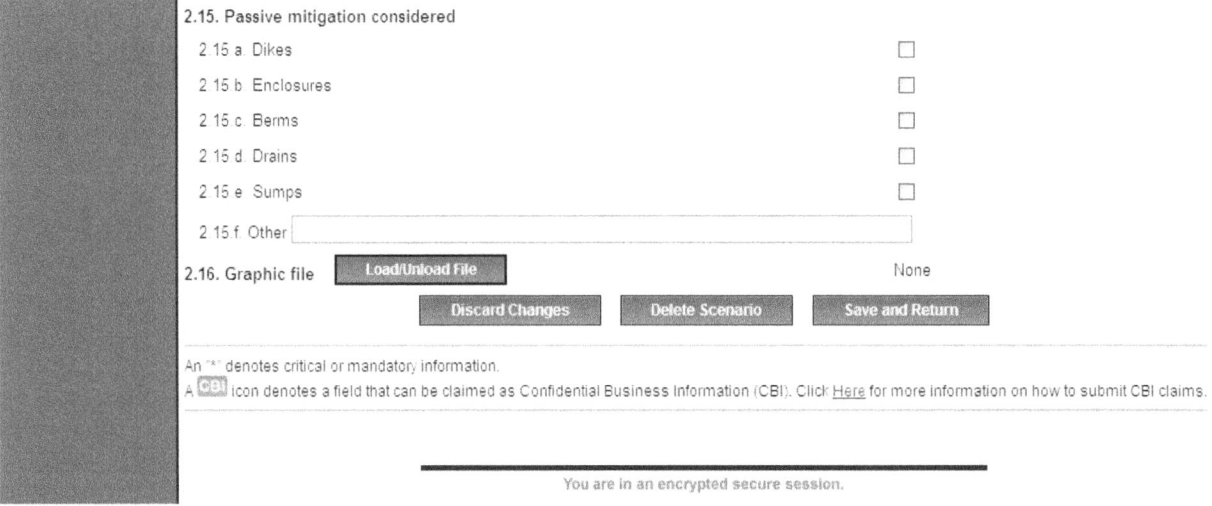

2.15   Passive mitigation considered:

Select one or more of the passive mitigations considered within distance to endpoint by checking the box that corresponds with the specified receptor in fields 2.15 a-f. This field is an indication that officially designated wildlife sanctuaries, preserves, or refuges are within a circle whose center is the point of the release and the radius is determined by the distance to the endpoint.

a. **Dikes:** A dike is a low wall that acts as a barrier to prevent a spill from spreading.

b. **Enclosures:** An enclosure is a physical containment of the release within a structure (e.g., a building).

c. **Berms:** A berm is a mound or wall of earth at the top or bottom of a slope that prevents a spill from spreading.

d. **Drains:** A drain is a channel that carries off surface water.

e. **Sumps:** A sump is a pit or tank that catches liquid runoff for drainage or disposal.

f. **Other:** List other types of passive mitigation.

2.16 Graphics file:

You may load one graphic file to illustrate each release scenario by clicking the "Load/Unload File" button. Entering a graphics file name in this field will not automatically cause that file to be included in your RMP submission. Graphics will be accepted in either GIF or JPEG file format. If you have a graphics file present in your RMP that you would like to remove, you can also delete the graphic file using the "Load/Unload File" button.

You can select one of the three buttons at the bottom of the page: "Discard Changes", "Delete Scenario", or "Save and Return".

Section 3. Toxics: Alternative Release

Alternative release scenario analysis of Program 2 and Program 3 processes as follows:

Present one alternative release scenario for **each** regulated toxic substance held above the threshold quantity in a Program 2 or 3 processes, including the substance considered in the worst-case analysis.

Present **one** alternative release scenario to represent **all** flammable substances held above the threshold quantity in a Program 2 or 3 processes.

Note that alternative release scenarios should be those that will reach an endpoint offsite, unless no such scenario exists.

You may include one graphic (map or diagram) in electronic format for each release scenario that you report, but it is not required.

Complete this section for each toxics alternative release scenario that you report.

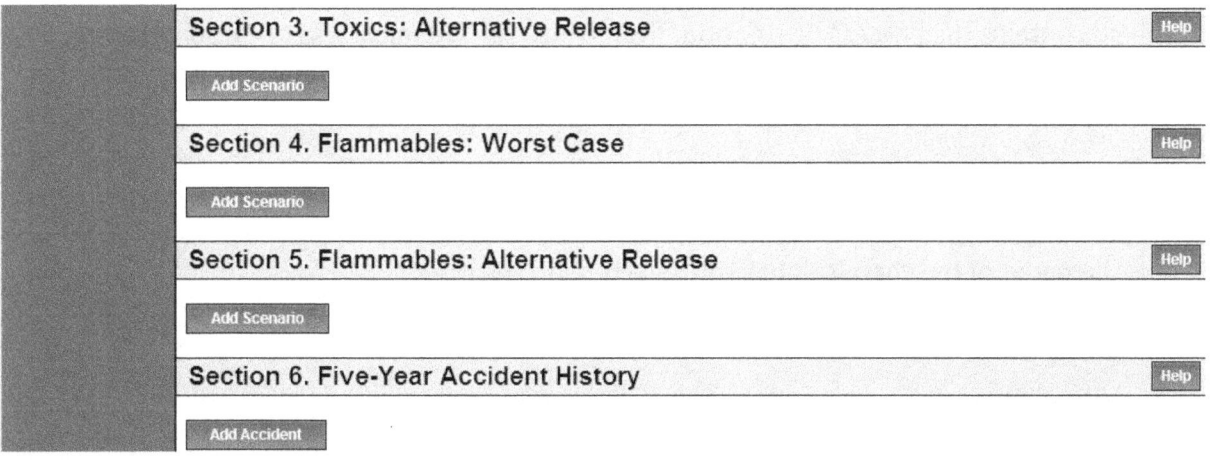

Select the "Add Scenario" button under *Section 3.Toxics: Alternative Release*

> NOTE: Please remember to periodically use the "Save and Return" button at the bottom of the page. As a security measure, if you remain inactive in the system for 19 minutes, the system will log you out at the 20th minute and your entered data will not be saved! Each time you save, the system will return you to the RMP*eSubmit Section Selection page. Click the "Enter Registration Information" button to enter more data.

> NOTE: If you do not wish to keep the information you've entered, click the "Discard Changes" button.

The following is a discussion of each element in *Section 3. Toxics: Alternative Release.*

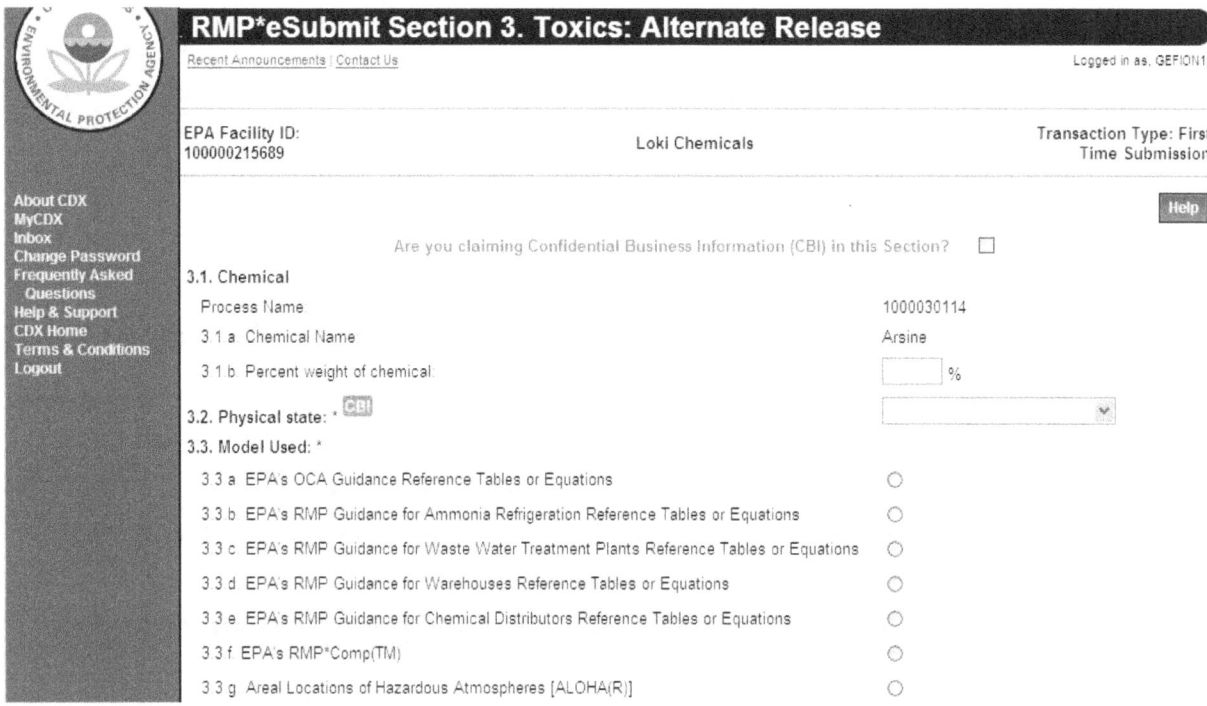

3.1   Chemical

a. Process name:

This field contains the process description, from which the chemical was entered in *Section 1. Processes.* This field is disabled and cannot be edited as it reflects information entered in *Section 1. Processes.*

b. Chemical name:

This field contains the regulated toxic chemical that you evaluated in the alternative scenario. This is the name of the chemical that you selected in 1.17.

c. Percent weight of chemical (if in a mixture):

If your alternative scenario involves the release of a mixture containing a regulated substance, enter the percentage weight of the regulated substance in the mixture. (Leave blank if it is not a mixture.)

3.2   Physical state:

Select the physical state of the chemical as it is released in the scenario from the drop down list.

- Gas (Select if the chemical is a gas)
- Liquid (Select if the chemical is a liquid)
- Gas liquified by pressure
- Gas liquified by refrigeration

3.3   Model used:

Select one of the options for the source of your results for your alternative release. If the *other model name* option is selected, you must specify the other model name in the *Model Used Text (if Other)* field. The options are list below:

a. EPA's *OCA Guidance* Reference Tables or Equations
b. EPA's RMP *Guidance for Ammonia Refrigeration* Reference Tables or Equations (toxics only) (now in *General Risk Management Program Guidance*)
c. EPA's *RMP Guidance for Waste Water Treatment Plants* Reference Tables or Equations (now in *General Risk Management Program Guidance*)
d. EPA's *RMP Guidance for Warehouses* Reference Tables or Equations
e. EPA's *RMP Guidance for Chemical Distributors* Reference Tables or Equations
f. EPA's *RMP*Comp™*
g. EPA's *Areal Locations of Hazardous Atmospheres (ALOHA®)*
h. Other model name (specify)

3.4. Scenario: *	Other ⌄
Scenario Text (if Other selected):	
3.5. Quantity released (lbs): * CBI	
3.6. Release rate (lbs/minute): * CBI	
3.7. Release duration (minutes): * CBI	
3.8. Wind speed (meters/second): *	
3.9. Atmospheric stability class: *	⌄
3.10. Topography: *	⌄
3.11. Distance to endpoint (miles): *	☐ Launch EPA's RMP*Comp
3.12. Estimated residential population within distance to endpoint (numbers): *	
3.13. Public receptors within distance to endpoint	
3.13.a. Schools	☐
3.13.b. Residences	☐
3.13.c. Hospitals	☐
3.13.d. Prison/Correctional facilities	☐
3.13.e. Recreational areas	☐
3.13.f. Major commercial, office or industrial areas	☐
3.13.g. Other	

3.4   Scenario:

Select one of the following that describes your alternative release scenario from the drop down list or enter another "Scenario Text":

- **Transfer Hose Failure:** Failure of the connection between two or more vessels. Liquid.
- **Pipe Leak:** Release through a rupture in a pipe
- **Vessel Leak:** Release through a rupture in a vessel
- **Overfilling:** Release due to filling a pipe, vessel, or other container past its capacity
- **Rupture Disk/Relief Valve:** Release due to failure of a rupture disk/relief valve to function properly. A rupture disk/relief valve is a valve that relieves pressure beyond a specified limit; a relief valve recloses upon return to normal operating conditions
- **Excess Flow Valve Failure:** Release caused by the failure of excess flow device to function properly and prevent surges from reaching downstream equipment
- **Other** (specify): Specify your scenario if not described in the drop-down menu

3.5   Quantity released (lbs):

Enter the quantity of toxic chemical that you used for your alternative scenario analysis in pounds in whole numbers. If you have less than one pound released, round up to one pound. You may want to clarify this in your *Executive Summary*.

3.6   Release rate (lbs/minute):

Enter the rate of release to the outside air in pounds per minute in whole numbers. For example:

> 4.3 pounds should be reported as... 4 pounds

19 pounds should be reported as... 20 pounds

See the General Risk Management Program Guidance for more information.

3.7   Release duration (minutes):

Enter the length of time in minutes (0.1 to 9999.9) for the release to the outside air of the quantity that you chose for the alternative scenario.

For gases, a gas liquified by pressurization alone, or a gas liquified by refrigeration where the released refrigerated liquid forms a pool of 1 cm or less in depth, you should assume that the release duration is 10 minutes.

For a liquid or a gas liquified by refrigeration where the released refrigerated liquid forms a pool deeper than 1 cm, the release duration should be the time required for a pool formed by the released substance to completely vaporize.

Although in some cases it may take longer than 60 minutes for the pool to completely volatilize, some dispersion models use the release rate and calculate the maximum downwind dispersion distance within 60 minutes. Therefore, you may enter 60 minutes for your duration, even if the duration from your modeling is longer than 60 minutes. You can also enter the exact duration from your modeling up to 9999.9 minutes.

See General Risk Management Program Guidance for more information.

3.8   Wind speed (meters/second):

If you used the OCA guidance or one of EPA's model program guidance documents, indicate 3 meters per second. If you modeled your scenario separately, provide the wind speed used.

3.9   Atmospheric stability class:

If you used the OCA Guidance or one of EPA's model program guidance documents, list D stability. If you modeled your scenario separately, enter the stability class used.

3.10   Topography (select one):

Select whether the local topography is urban or rural from the drop down list. Urban means that there are many obstacles in the immediate area; obstacles include buildings or trees. Rural means that the terrain is generally flat and unobstructed in the immediate area.

3.11   Distance to endpoint (miles):

Enter the distance to the endpoint in miles to 2 significant digits, using the endpoint specified for the chemical in 40 CFR part 68, Appendix A. Convert your modeling results into miles by dividing the distance in feet by 5280 or yards by 1760. Refer to the following to determine 2 significant digits:

0.397 miles should be reported as 0.40 miles

9.345 miles should be reported as 9.3 miles

20.764 miles should be reported as 21.0 miles

3.12   Estimated residential population within distance to endpoint (numbers):

Enter the population within the circle with a center at the point of the release and a radius determined by the distance to the endpoint to two significant digits (e.g., 5,500 people rather than 5,483). Population estimates include only residential populations.

3.13   Public receptors within distance to endpoint:

Select one or more of the public receptors. Public receptors must be identified within the circle with a center at the point of the release and a radius determined by the distance to the endpoint. Public receptor means locations where members of the public may be exposed to toxic concentrations, radiant heat, or overpressure as a result of the release. Public receptors include locations within the facility's property boundary to which the public has routine and unrestricted access during or outside business hours (e.g., a recreation field). Locations inhabited or occupied by the public at any time without restriction by the source (such as fences or security guards) are public receptors (see the *General Guidance for Risk Management Programs* for more information on identifying public receptors). You do not need to list specific locations or estimate populations at these locations. The presence of these receptors may be determined using local street maps. Select all that apply in fields 3.13 a-g.

  a. **Schools:** Public and private elementary, secondary, or post-secondary educational institutions (e.g., colleges).
  b. **Residences:** Public and private residences and dwellings wherever people live.
  c. **Hospitals:** Public and Private Hospitals. Places that provide emergency care and or long term care of the sick or elderly.
  d. **Prisons/Correction facilities:** Prisons where men or women are incarcerated. Holding places or state of confinement for criminals.
  e. **Recreation areas:** Recreation areas include stadiums, parks, and public pools.
  f. **Major commercial, office industrial areas:** Commercial, office, or industrial areas include industrial parks, office buildings, shopping malls, commercial areas, and commercial farms.
  g. **Other:** Include any other additional information here.

3.14. Environmental receptors within distance to endpoint

3.14 a. National or State parks, forests, monuments   ☐

3.14 b. Officially designated wildlife sanctuaries, preserves or refuges   ☐

3.14 c. Federal wilderness area   ☐

3.14 d. Other [ ]

3.15. Passive mitigation considered

3.15 a. Dikes   ☐

3.15 b. Enclosures   ☐

3.15 c. Berms   ☐

3.15 d. Drains   ☐

3.15 e. Sumps   ☐

3.15 f. Other [ ]

3.14   *Environmental receptors within distance to endpoint:*

Select one or more environmental receptors. Environmental receptors must be identified within the circle with a center at the point of the release and a radius determined by the distance to the endpoint by clicking the check box in field 3.14 a-d. Environmental receptor means natural areas, such as national or state parks, forests, or monuments; officially designated wildlife sanctuaries, preserves, refuges, or areas; and federal wilderness areas that could be exposed at any time to toxic concentrations, radiant heat, or overpressure as a result of the release. Environmental receptors can be identified on local U.S. Geological Survey maps, which can be found at many libraries and online via the USGS Maps, Imagery, and Publications Website. Select all that apply:

a. National or state parks, forests, or monuments
b. Officially designated wildlife sanctuaries, preserves, refuges, or areas
c. Federal wilderness areas
d. Other (Specify) (Include any other additional information.)

3.15   *Passive mitigation considered:*

Mitigation means specific activities, technologies, or equipment designed or deployed to capture or control substances that have been released to minimize exposure of the public or the environment. Passive mitigation means equipment, devices, or technologies that function without human, mechanical, or other energy input. Select all that were considered in defining the release quantity or rate to the worst-case scenario or alternative release scenario.

a. **Dike:** A dike is a low wall that acts as a barrier to prevent a spill from spreading.
b. **Enclosures:** An enclosure is a physical containment of the release within a structure [e.g., a building].
c. **Berm:** A berm is a mound or wall of earth at the top or bottom of a slope that prevents a spill from spreading.
d. **Drain:** A drain is a channel that carries off surface water.
e. **Sump:** A sump is a pit or tank that catches liquid runoff for drainage or disposal.
f. **Other:** (specify)

3.16   Active mitigation considered:

Active mitigation means equipment, devices, or technologies that need human, mechanical, or other energy input to function. Select all that were considered in defining the release quantity or rate of the alternative release scenario.

    a. **Sprinkler Systems:** A system for protecting a building against fire by means of overhead pipes which convey an extinguishing fluid through heat activated outlets.

    b. **Deluge Systems:** A system to overflow an area of a release with water or other extinguishing fluid.

    c. **Water Curtain:** A spray of water from a horizontal pipe through nozzles: the curtain may be activated manually or automatically.

    d. **Neutralization:** A means of making a toxic chemical harmless through chemical reaction.

    e. **Excess Flow Valve:** A device in the outlet of a vessel at a hose connection that stops the flow or liquid or gas if the piping or hoses downstream fail and a predetermined excess flow rate is reached.

    f. **Flares:** A device for disposing of combustible gases from a chemical process by burning them in the open.

    g. **Scrubbers:** A pre-release protection measure that uses water or aqueous mixtures containing scrubbing reagents to remove discharging liquids and possibly also treating the discharging chemical.

    h. **Emergency shutdown systems:** Controls that are triggered when process limits are exceeded and that shutdown process.

    i. **Other (specify):** Enter a type of passive mitigation considered other than what is listed above.

3.17   Graphics file:

You may load one graphic file to illustrate each release scenario by clicking on the "Load/Unload File" button. Entering a graphics file name in this field will not automatically cause that file to be included in your RMP submission. Graphics will be accepted in either GIF or

JPEG file format. If you have a graphics file present in your RMP that you would like to remove, you can also unload the graphic file using the "Load/Unload File" button.

Section 4. Flammables: Worst Case

Report one worst-case release scenario for each Program 1 process. Program 1 processes must have no public receptors within the distance to the endpoint in the worst-case analysis.

If your facility has Program 2 or Program 3 processes, report <u>one</u> worst-case release scenario to represent **all** Program 2 and Program 3 processes having toxic regulated substances present above the threshold quantity, **and** <u>one</u> worst-case release scenario to represent **all** Program 2 and Program 3 processes having flammable regulated substances present above the threshold quantity. If you have more than one Program 2 or 3 process, you will report the worst-case release scenario for the Program 2 or 3 process that would have the greatest potential impact on the public (i.e., the greatest distance to endpoint). You may also need to submit an additional worst-case scenario for either hazard class (i.e., toxic or flammable), if a worst-case release from elsewhere at your facility would potentially affect a different set of public receptors than those affected by your initial worst-case scenario(s).

Complete this section for each flammable worst-case scenario you report.

The following is a discussion of each element in *Section 4. Flammables: Worst Case.*

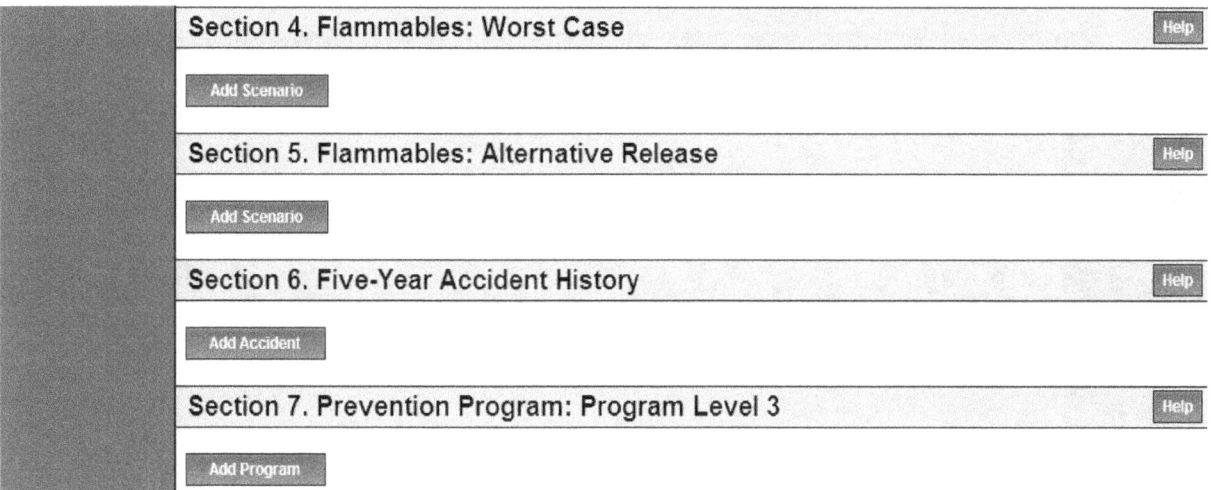

Select the "Add Scenario" button under *Section 4.Flammables: Worst Case*

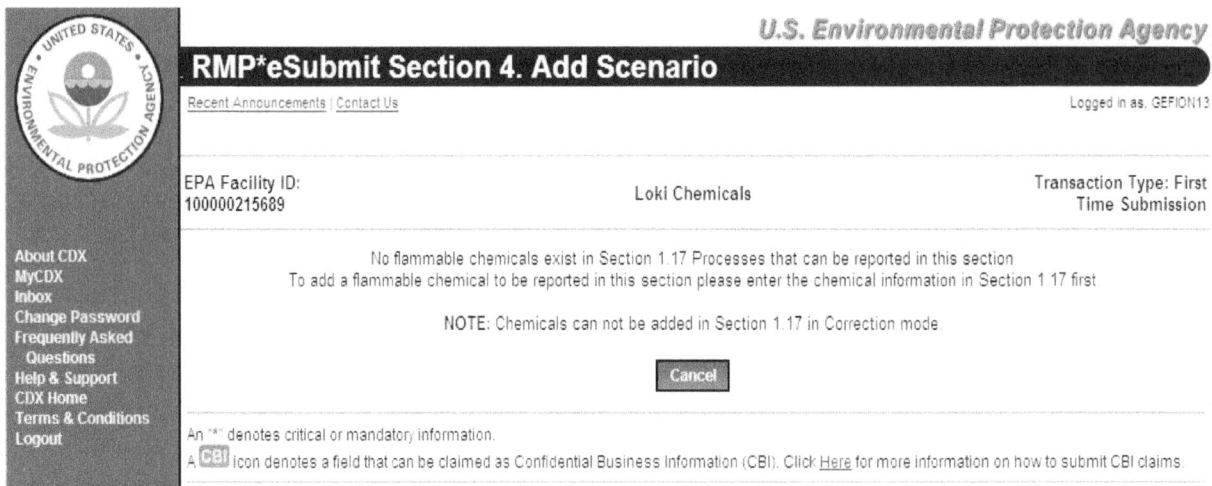

NOTE: Please remember to periodically use the "Save and Return" button at the bottom of the page. As a security measure, if you remain inactive in the system for 19 minutes, the system will log you out at the 20th minute and your entered data will not be saved! Each time you save, the system will return you to the RMP*eSubmit Section Selection page. Click the "Enter Registration Information" button to enter more data.

NOTE: If you do not wish to keep the information you've entered, click the "Discard Changes" button.

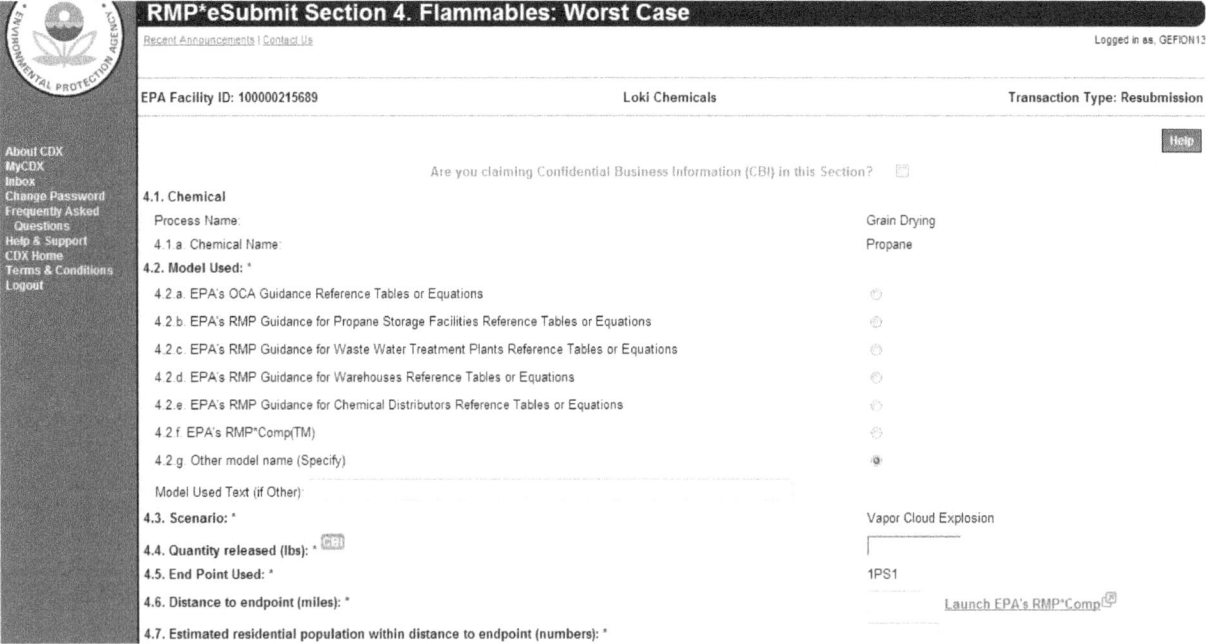

4.1   Chemical

a. Process name:

The process description from which the chemical was entered in *Processes 1.17*. This field is disabled and cannot be edited as it reflects information entered in *Section 1. Processes*.

b. Chemical name:

The regulated flammable chemical you evaluated in the worst-case scenario.

4.2   Model used (select one):

Select one of the options for the source of your results for your worst-case release:

   a. EPA's *OCA Guidance* Reference Tables or Equations
   b. EPA's RMP *Guidance for Propane Storage Facilities* Reference Tables or Equations (Flammable only)
   c. EPA's RMP *Guidance for Waste Water Treatment Plants* Reference Tables or Equations (now in *General Risk Management Program Guidance*)
   d. EPA's RMP *Guidance for Warehouses* Reference Tables or Equations
   e. EPA's RMP *Guidance for Chemical Distributors* Reference Tables or Equations
   f. EPA's *RMP* Comp™*
   g. Other model name (specify)

4.3   Scenario:

This data element is fixed. By regulation, for flammables, the worst case assumes an instantaneous release and a **vapor cloud explosion**, which is an explosion of a cloud containing a flammable vapor or gas and air.

4.4   Quantity released (lbs):

Enter the quantity of the flammable substance you used for your worst case scenario analysis in pounds to two significant digits. If you have less than one pound released, round up to one pound. You may want to clarify this in your *Executive Summary.*

4.5   Endpoint used:

This data element is fixed. Because the scenario is fixed by regulation as vapor cloud explosions, the endpoint which applies to vapor cloud explosions is fixed at **1 psi** overpressure.

4.6   Distance to endpoint (miles):

Enter the distance to the endpoint in miles to two significant digits, using the endpoint specified for the chemical in 40 CFR part 68, Appendix A. Convert your modeling results into miles by dividing the distance in feet by 5280 or yards by 1760. Refer to the following to determine two significant digits:

   0.397 miles should be reported as 0.40 miles

   9.345 miles should be reported as 9.3 miles

   20.764 miles should be reported as 21.0 miles

4.7   Residential population within distance to endpoint:

Estimate the population within the circle with a center at the point of the release and a radius determined by the distance to the endpoint to two significant digits (e.g., 5,500 people rather than 5,483). Population estimates include only residential populations.

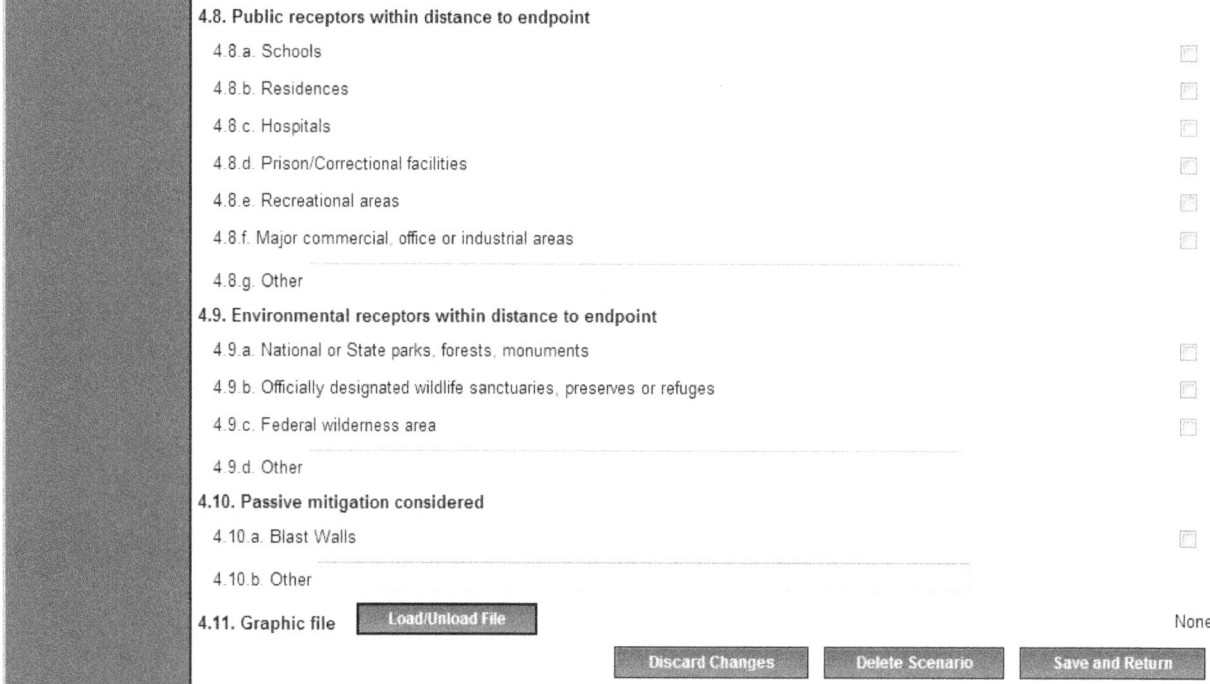

4.8   Public receptors within distance to endpoint:

Public receptors must be identified within the circle with a center at the point of the release and a radius determined by the distance to the endpoint. Public receptor means locations where members of the public may be exposed to toxic concentrations, radiant heat, or overpressure as a result of the release. Public receptors include locations within the facility's property boundary to which the public has routine and unrestricted access during or outside business hours (e.g., a recreation field). Locations inhabited or occupied by the public at any time without restriction by the source (such as fences or security guards) are public receptors (see the *General Risk Management Program Guidance* for more information on identifying public receptors). You do not need to list specific locations or estimate populations at these locations. The presence of these receptors may be determined using local street maps.

Select all that apply.

   a. **Schools:** Public and private elementary, secondary, or post-secondary educational institutions (e.g., colleges).
   b. **Residences:** Public and private residences and dwellings wherever people live.
   c. **Hospitals:** Public and private hospitals. Places that provide emergency care and or long term care for the sick and elderly.
   d. **Prisons/Correctional facilities:** Prisons where men or women are incarcerated. Holding places or state of confinement for criminals.

e. **Recreation areas:** Recreation areas include stadiums, parks, and public pools.

f. **Major commercial, office, or industrial areas:** Commercial, office, or industrial areas include industrial parks, office buildings, shopping malls, commercial areas, and commercial farms.

g. **Other:** Include any other additional information here.

4.9   *Environmental receptors within distance to endpoint:*

Environmental receptors must be identified within the circle with a center at the point of the release and a radius determined by the distance to the endpoint. Environmental receptors means areas, such as national or state parks, forests, or monuments; officially designated wildlife sanctuaries, preserves, refuges, or areas; and federal wilderness areas that could be exposed at any time to toxic concentrations, radiant heat, or overpressure as a result of the release. Environmental receptors can be identified on local U.S. Geological Survey maps, which can be found at many libraries and online via the USGS Website. Select all that apply.

a. National or state parks, forests, or monuments
b. Officially designated wildlife sanctuaries, preserves, or refuges
c. Federal wilderness areas
d. Other (specify) (Include any other additional information here)

4.10   *Passive mitigation considered:*

Mitigation means specific activities, technologies, or equipment designed or deployed to capture or control substances that have been released to minimize exposure of the public or the environment. Passive mitigation means equipment, devices, or technologies that function without human, mechanical, or other energy input. Select all that were considered in defining the release quantity or rate to the worst-case scenario or alternative release scenario. If the selection for field 4.10.a. does not apply, use the text field in 4.10.b. to enter the passive mitigation considered in your scenario.

a. **Blast Wall:** A heavy wall used to isolate buildings or areas that contain highly combustible or explosive materials
b. **Other** (specify)

4.11   *Graphics file name:*

You may load one graphic file to illustrate each release scenario by clicking on the "Load/Unload File" button. Entering a graphics file name in this field will not automatically cause that file to be included in your RMP submission. Graphics will be accepted in either GIF or JPEG file format. If you have a graphics file present in your RMP that you would like to remove, you can also unload the graphic file using the "Load/Unload File" button.

Section 5. Flammables: Alternative Release

Complete this section for each flammable alternative release scenario you report. If a flammable substance is used in multiple processes, only one scenario is required. You will only be able to add or update a scenario if you have entered a flammable chemical in *Section 1. Processes* section.

Complete this section for each flammable alternative release scenario you report.

The following is a discussion of each element in *Section 5. Flammables: Alternative Release.*

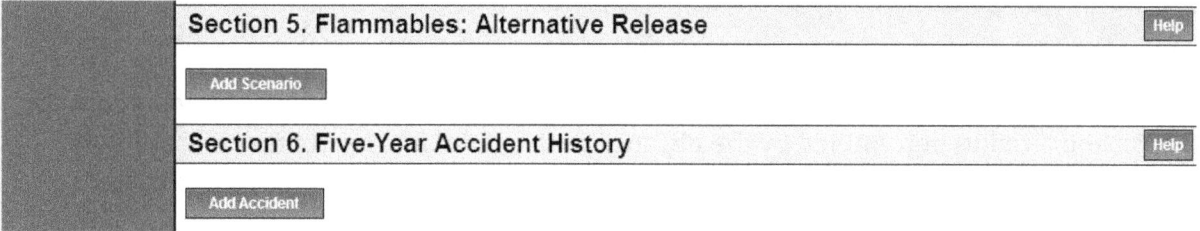

Select the "Add Scenario" button under Section 5.Flammables: Alternative Release

> NOTE: *Please remember to periodically use the "Save and Return" button at the bottom of the page. As a security measure, if you remain inactive in the system for 19 minutes, the system will log you out at the 20th minute and your entered data will not be saved! Each time you save, the system will return you to the RMP*eSubmit Section Selection page. Click the "Enter Registration Information" button to enter more data.*

> NOTE: *If you do not wish to keep the information you've entered, click the "Discard Changes" button.*

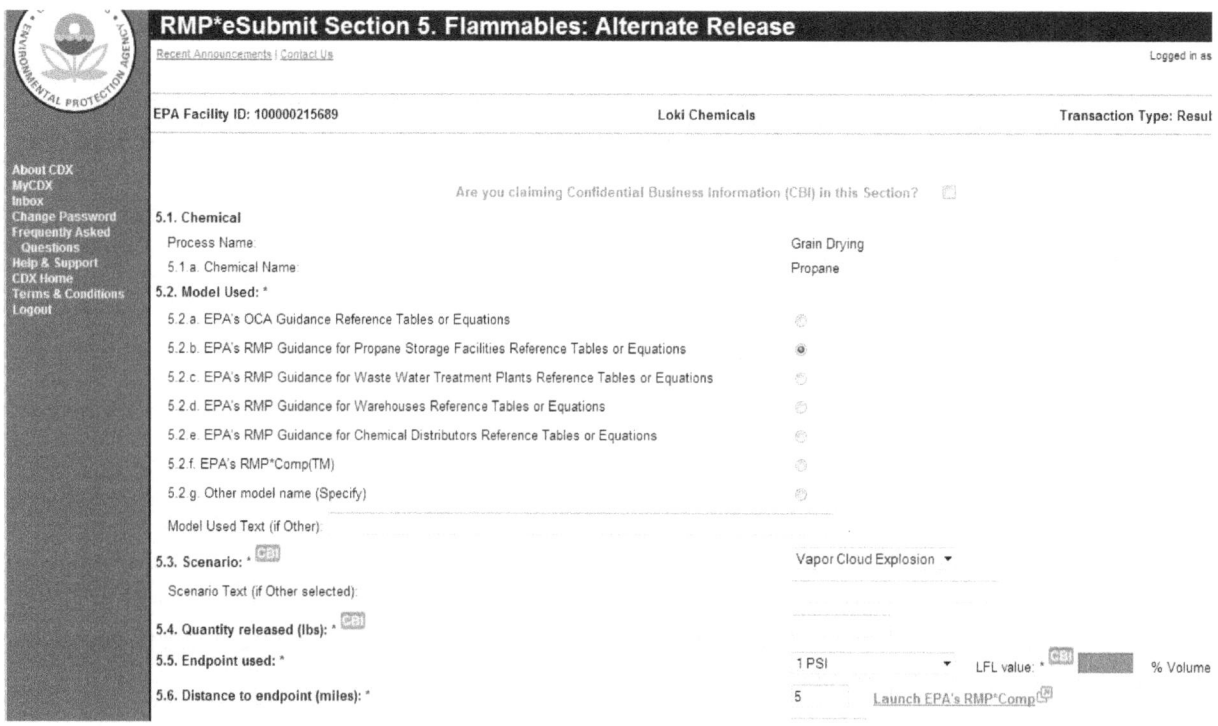

5.1   *Chemical*

a. Process name:

The process description from which the chemical was entered in *Processes 1.17.*

5.2   *Model used:*

Select the source of your results for your alternative release from the options:

a.   EPA's *OCA Guidance* Reference Tables or Equations
b.   EPA's RMP *Guidance for Propane Storage Facilities* Reference Tables or Equations (Flammables only)
c.   EPA's RMP *Guidance for Waste Water Treatment Plants* Reference Tables or Equations (now in *General Risk Management Program Guidance*)
d.   EPA's RMP *Guidance for Warehouses* Reference Tables or Equations
e.   EPA's RMP *Guidance for Chemical Distributors* Reference Tables or Equations
f.   EPA's *RMP*Comp™*
g.   Other model name (specify)

5.3   *Scenario:*

Select one of the following or enter another scenario in *Other:*

- **Vapor Cloud Explosion:** An explosion of a cloud containing a flammable vapor or gas and air.
- **Fireball**: The atmospheric burning of a fuel-air cloud in which the energy is mostly emitted in the form of radiant heat. As the buoyancy forces of the hot gases begin to dominate, the burning cloud rises and becomes spherical in shape. Fireballs are often caused by the ignition of a vapor cloud of a flammable substance.
- **BLEVE**: Boiling Liquid Expanding Vapor Explosion (BLEVE) is used to describe the sudden rupture of a vessel/system containing liquified flammable gas under pressure due to radiant heat flux. The pressure burst and the flashing of the liquid to vapor creates a blast wall and potential missile damage, and immediate ignition of the expanding fuel-air mixture leads to an intense combustion creating a fireball.
- **Pool Fire**: The combustion of material evaporating from a layer of liquid at the base of the fire.
- **Jet Fire**: Gas or liquid discharging or venting from a rupture will form a gas jet that "blows" into the atmosphere in the direction the hole is facing, all the while entraining and mixing with air. If the jet is flammable and encounters an ignition source, a flame jet may form.
- **Vapor Cloud Fire**: A flash fire results from the ignition of a released flammable cloud in which there is essentially no increase in the combustion rate.
- **Other**: (specify)

5.4   Quantity released (lbs):

Enter the quantity of the flammable substance you used for your alternative scenario analysis in pounds to two significant digits. If you have less than 1 pound released, round up to 1 pound. You may want to clarify that in your *Executive Summary.*

5.5   Endpoint used:

For vapor cloud explosions, the endpoint is 1 psi overpressure; for a fireball the endpoint is 5 kilowatts per square meter for 40 seconds; for vapor cloud fires or jet fires, a lower flammability limit (expressed as a percentage) may be listed as specified in NFPA documents or other generally recognized sources. These are listed in the OCA Guidance. Enter the endpoint used in the text field.

5.6   Distance to endpoint (miles):

Enter the distance to the endpoint in miles to two significant digits, using the endpoint specified for the chemical in 40 CFR part 68, Appendix A. Convert your modeling results into miles by dividing the distance in feet by 5280 or yards by 1760. Refer to the following to determine two significant digits:

> 0.397 miles should be reported as 0.40 miles
>
> 9.345 miles should be reported as 9.3 miles
>
> 20.764 miles should be reported as 21.0 miles

5.7   Residential population within distance to endpoint:

Estimate the population within the circle with a center at the point of the release and a radius determined by the distance to the endpoint to two significant digits (e.g., 5,500 people rather than 5,483). Population estimates include only residential populations.

5.8. Public receptors within distance to endpoint

5.8.a. Schools	☐
5.8.b. Residences	☑
5.8.c. Hospitals	☐
5.8.d. Prison/Correctional facilities	☐
5.8.e. Recreational areas	☑
5.8.f. Major commercial, office or industrial areas	☐

5.8.g. Other

5.9. Environmental receptors within distance to endpoint

5.9.a. National or State parks, forests, monuments	☐
5.9.b. Officially designated wildlife sanctuaries, preserves or refuges	☑
5.9.c. Federal wilderness area	☐

5.9.d. Other

5.10. Passive mitigation considered

5.10.a. Dikes	☑
5.10.b. Fire walls	☐
5.10.c. Blast walls	☐
5.10.d. Enclosures	☐

5.10.e. Other

5.11. Active mitigation considered

5.11.a. Sprinkler systems	☐
5.11.b. Deluge systems	☐
5.11.c. Water curtain	☐
5.11.d. Excess flow valve	☑

5.11.e. Other

5.12. Graphic file   [Load/Unload File]   None

[Discard Changes]   [Delete Scenario]   [Save and Return]

5.8   *Public receptors within distance to endpoint:*

Public receptors must be identified within the circle with a center at the point of the release and a radius determined by the distance to the endpoint. Public receptor means locations where members of the public may be exposed to toxic concentrations, radiant heat, or overpressure as a result of the release. Public receptors include locations within the facility's property boundary to which the public has routine and unrestricted access during or outside business hours (e.g., a recreation field). Locations inhabited or occupied by the public at any time without restriction by the source (such as fences or security guards) are public receptors (see the *General Risk Management Program Guidance* for more information on identifying public receptors). You do not need to list specific locations or estimate populations at these locations. The presence of these receptors may be determined using local street maps. Select all that apply:

a. **Schools:** Public and private elementary, secondary, or post-secondary educational institutions (e.g., colleges).
b. **Residences:** Private residences and dwellings wherever people live.
c. **Hospitals:** Health care facilities.

d. **Prisons or Correctional facilities:** Places of incarceration.

e. **Recreation areas:** Including stadiums, parks, and public pools.

f. **Major commercial, office, or industrial areas:** Including industrial parks, office buildings, shopping malls, commercial areas, and commercial farms.

g. **Other:** (Include any other additional information here)

5.9   *Environmental receptors within distance to endpoint:*

Environmental receptors must be identified within the circle with a center at the point of the release and a radius determined by the distance to the endpoint. Environmental receptor means natural areas, such as national or state parks, forests, or monuments; officially designated wildlife sanctuaries, preserves, refuges, or areas; and federal wilderness areas that could be exposed at any time to toxic concentrations, radiant heat, or overpressure as a result of the release. Environmental receptors can be identified on local U.S. Geological Survey maps, which can be found at many libraries and online via the USGS Website. Select all that apply by clicking the appropriate check boxes.

a. National or state parks, forests, or monuments

b. Officially designated wildlife sanctuaries, preserves, or refuges

c. Federal wilderness areas

d. Other (Include any other additional information here)

5.10   *Passive mitigation considered:*

Mitigation means specific activities, technologies, or equipment designed or deployed to capture or control substances that have been released to minimize exposure of the public or the environment. Passive mitigation means equipment, devices, or technologies that function without human, mechanical, or other energy input. Select all that were considered in defining the release quantity or rate to the worst-case scenario or alternative release scenario by clicking the appropriate check boxes.

a. **Dike:** A dike is a low wall that acts as a barrier to prevent a spill from spreading.

b. **Fire Wall:** A wall constructed to prevent the spread of fire.

c. **Blast Wall:** A heavy wall used to isolate buildings or areas that contain highly combustible or explosive materials.

d. **Enclosure:** Physical containment of the release within a structure (e.g., a building).

e. **Other:** (specify)

5.11   *Active mitigation considered:*

Active mitigation means equipment, devices, or technologies that need human, mechanical, or other energy input to function. Select all that were considered in defining the release quantity or rate of the alternative release scenario.

a. **Sprinkler Systems:** A system for protecting a building against fire by means of overhead pipes which convey an extinguishing fluid through heat activated outlets.

b. **Deluge Systems:** A system to overflow an area of a release with water or other extinguishing fluid.

c. **Water Curtain:** A spray of water from a horizontal pipe through nozzles, the curtain may be activated manually or automatically.

d. **Excess Flow Valve**: A system for diverting overflow.

e. **Other** (specify):

5.12 *Graphics file name:*

You may load one graphic file to illustrate each release scenario by clicking on the "Load/Unload File" button. Entering a graphics file name in this field will not automatically cause that file to be included in your RMP submission. Graphics will be accepted in either GIF or JPEG file format. If you have a graphics file present in your RMP that you would like to remove, you can unload the graphic file using the "Load/Unload" File button.

Section 6: Five-Year Accident History

Complete this section for each reportable accident.

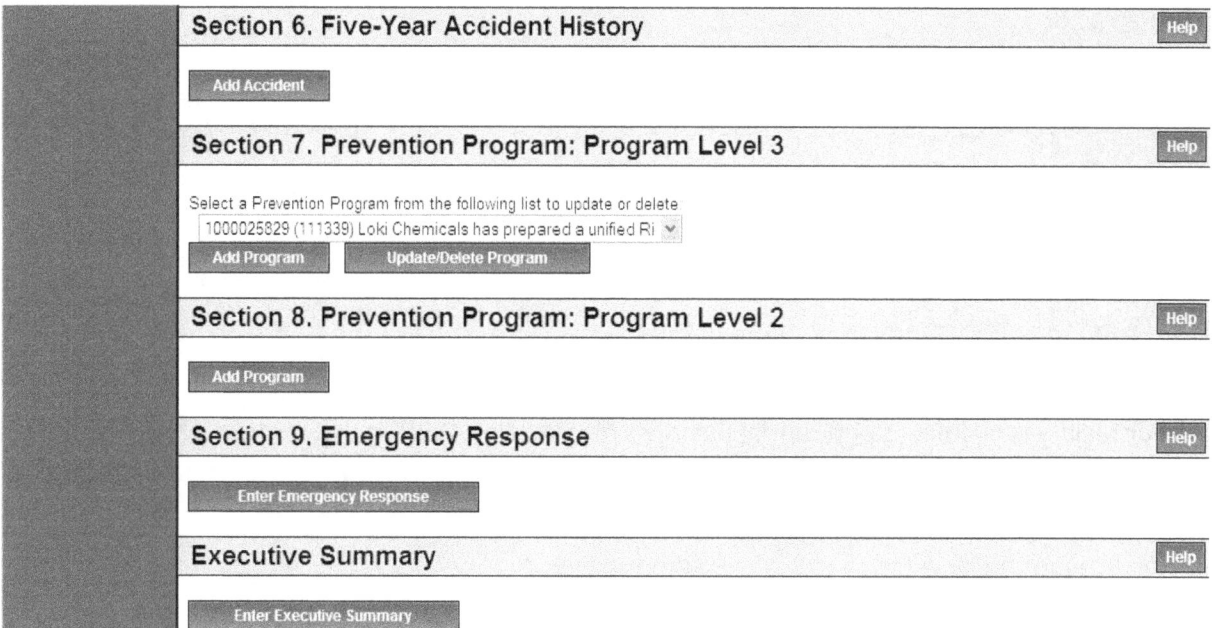

Select the "Add Accident" button under *Section 6. Five-Year Accident History*

> *NOTE: Please remember to periodically use the "Save and Return" button at the bottom of the page. As a security measure, if you remain inactive in the system for 19 minutes, the system will log you out at the 20th minute and your entered data will not be saved! Each time you save, the system will return you to the RMP*eSubmit Section Selection page. Click the "Enter Registration Information" button to enter more data.*

> *NOTE: If you do not wish to keep the information you've entered, click the "Discard Changes" button.*

The following is a discussion of each element in *Section 6. Five-Year Accident History.*

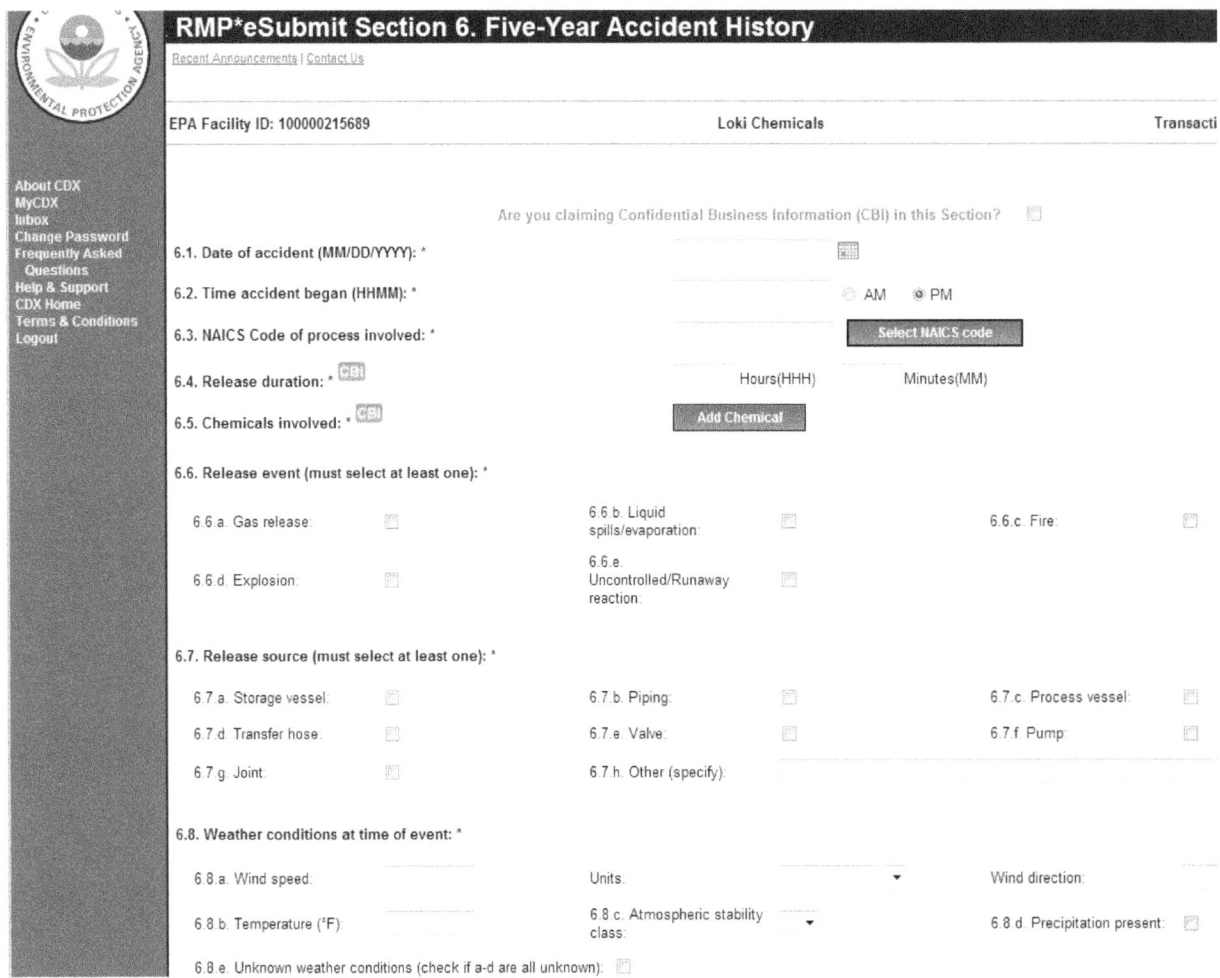

If your facility is claiming Confidential Business Information (CBI) in this section, check the box. If claiming CBI and you've checked the box, a pop-up message will appear describing the step that must be taken by the Preparer.

6.1   Date of accident:

Enter the date on which the accident occurred (MM/DD/YYYY).

6.2   Time accident began:

Enter the time the release began. Use standard time references (i.e., HHMM = 1237). You must also select the **AM** or **PM** check box option to indicate what time of day the accident occurred.

6.3   NAICS Code of process involved:

Provide an NAICS code by using the NAICS Code Selector field to select the North American Industry Classification System (NAICS) codes associated with your covered processes.

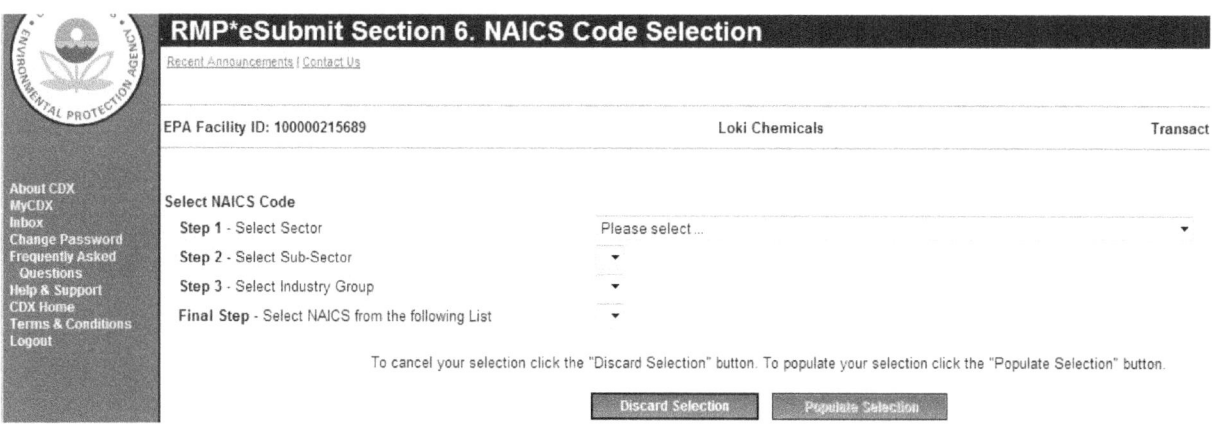

NAICS Code Selection: The North American Industry Classification System (NAICS) categorizes businesses by fitting them into descriptive categories that correspond to five-digit or six-digit codes. NAICS codes have replaced SIC codes, which you may be familiar with. For this data element you should provide the NAICS code that most closely corresponds to the process; it will not necessarily be the same NAICS code for you facility as a whole. You may also enter additional NAICS codes if you wish to identify other aspects of a process not captured by the NAICS codes for the primary activity.

You should determine the NAICS codes for your processes based on your activities on site using the 2012 North American Industry Classification System Manual, which can be viewed online: www.census.gov/epcd/www/naics.html.

Once you've selected the sector that most appropriately reflects the sector for your facility (Step 1), you must select the sub-sector, which enables you to select a more specific sector within your industry (Step 2). Next, you must select the industry group that represents your facility (Step 3). The last option is to select the NAICS code that reflects your facility (Final Step). All steps must be competed in succession. See below for more information.

> **Step 1 – Select Sector** *drop down list enables you to add the sector of the NAICS code to an accident scenario.*

> **Step 2 – Select Sub-Sector** *drop down list enables you to add the subsector of the NAICS code to an accident scenario.*

> **Step 3 – Select Industry Group** *drop down list enables you to add the industry group of the NAICS code to an accident scenario.*

> **Final Step – Select NAICS from following list** *drop down list enables you to select your NAICS code to an accident scenario.*

The "Populate Selection" button enables you to populate the five-year accident history record with any selections you've made from any of the available drop down list in the and returns you to the previous screen, *Section 6. Five-Year Accident History.*

The "Discard Selection" button enables you to discard any selections you've made from any of the available drop down list without saving and returns you to the previous screen, *Section 6. Five-Year Accident History.*

6.4   Release duration:

Enter the approximate length of time of the release in hours (Format: HHH) and in minutes (Format: MM).

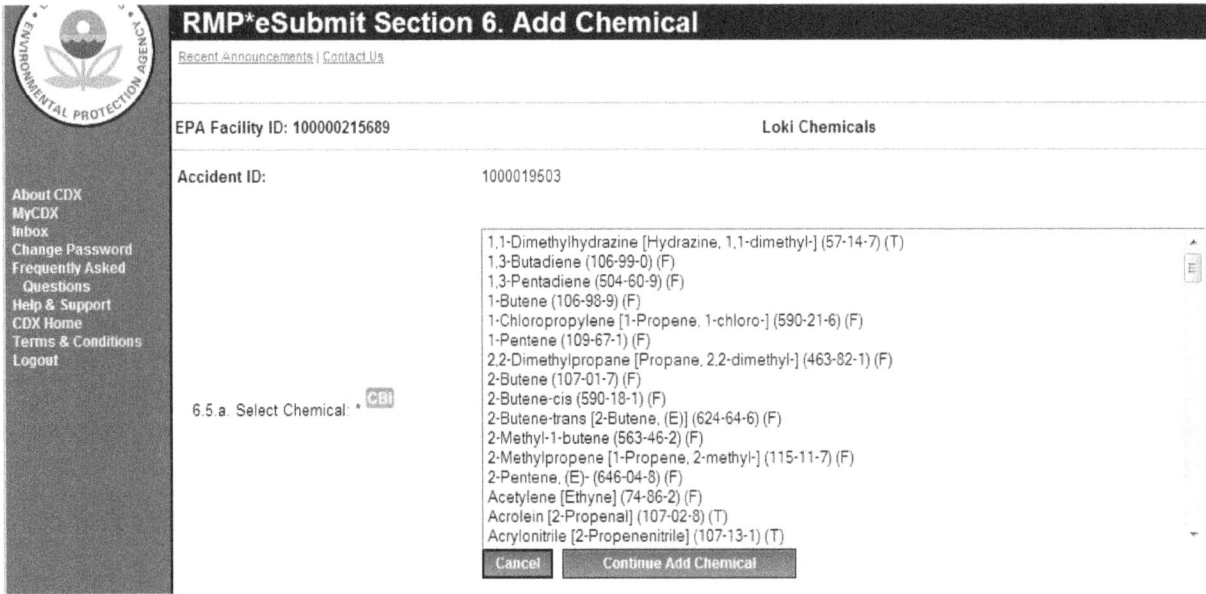

6.5   Chemicals Involved:

Indicate the regulated substance(s) released. Use the name of the substance as listed in Section 68.130 rather than a synonym. If the release was a NFPA-4 flammable mixture containing regulated flammables, you may list it as a "flammable mixture" and list all of the regulated substances contained in the mixture. For the quantity released, you only report the quantity of the entire mixture, not the individual substances. Only report chemicals that are listed substances.

   a. **Select Chemical:** Select chemical from the scrolling list and click the "Continue Add Chemical" button. Chemicals can be added one at a time.

   b. **Quantity released (lbs):** Provide the chemical quantity for your process chemical in this field by clicking the "Add Chemical" button. For each chemical reported in 1.17.c.1, report the maximum quantity (in pounds) held in the covered process at any one time during the calendar year to two significant digits. For example:

     5,333 pounds should be reported as 5,300 pounds
     107,899 pounds should be reported as 110,000 pounds
     128,000 pounds should be reported as 130,000 pounds

6.6   Release event:

Indicate which of the following release events best describes your accident. Select at least one item from the available check boxes.

   a. **Gas Release:** A gas release is a release of the substance in a vapor state. If you hold a gas liquified under refrigeration, report the release as a liquid spill.

c. **Liquid Spill/Evaporation:** A liquid spill/evaporation is a release of the substance in a liquid state with subsequent vaporization.

d. **Fire:** A product (e.g., fuel) in a state of combustion.

e. **Explosion:** A rapid chemical reaction with the production of noise, heat, and violent expansion of gasses.

f. **Uncontrolled/Runaway Reaction:** An indication that the release event involved an uncontrolled or runaway reaction. A release event caused by an uncontrolled chemical reaction that generates excessive heat, pressure, or harmful reaction products. Such events may involve highly exothermic chemical reactions, self-reactive substances (e.g., substances that undergo polymerization), unstable, explosive, or spontaneously combustible substances, substances that react strongly with water or other contaminants, oxidizers, peroxide-forming substances, or other types of chemical reactions that generate harmful products or byproducts. This category of release event may often occur in conjunction with one of the previous categories. In such cases, be sure to check this category in addition to any other applicable release event category (e.g., explosion). The burning of ordinary flammable substances is not typically included in this category.

6.7   Release source:

Select at least one:

a. **Storage Vessel:** A storage vessel is a container used only to store or hold (as opposed to react, mix, or move) a regulated substance. Storage vessels include transportation containers (e.g., railcars or tank trucks) being used for on-site storage.

b. **Piping:** Piping refers to a system of pipes used to carry a fluid or gas.

c. **Process Vessel:** A process vessel is a container in which regulated substances are manufactured, reacted, or mixed.

d. **Transfer Hose:** A transfer hose is a flexible tube used to connect, often temporarily, two or more vessels.

e. **Valve:** A valve is a device used to regulate the flow in piping systems or machinery. Relief valves open to release pressure in vessels.

f. **Pump:** A pump is a device that raises, transfers, or compresses fluids or that attenuates gases by suction or pressure or both.

g. **Joint:** The surface at which two or more mechanical components are united.

h. **Other (specify):**

6.8   Weather conditions at time of event:

This information is important to those concerned with modeling the effects of accidents. Reliable information from those involved in the incident or from an onsite weather station is ideal. However, the rule does not require your facility to have an on-site weather station. If you do not have an on-site weather station, use information from your local weather station, airport, or other source of meteorological data. Historical wind speed and temperature data (but not stability data), can be obtained from the National Climatic Data Center (NCDC) at 828-271-4800 or via the NCDC Website. NCDC staff can also provide information on the nearest weather station. To the extent possible, complete the following:

a.
  i. **Wind speed:** Wind speed is an estimate of how fast the wind is traveling.

  ii. **Wind speed units:** Indicate the units in which the speed is expressed as either miles per hour, meters per second, or knots.

  iii. **Wind direction:** Wind direction is the direction from which the wind comes. For example, a wind that blows from east to west would be described as having an eastern wind direction. Describe wind direction as one of the 16 standard compass readings [N, S, E, W, NE, SE, NW, SW, NNE, ENE, ESE, SSE, SSW, WSW, WNW, NNW], using the abbreviation. For example, wind direction must be reported as S for south, NE for northeast, or SSW for south-southwest.

b. **Temperature:** Enter the ambient temperature at the scene of the accident in degrees Fahrenheit. If you did not keep a record, you can use the high (for daytime releases) or low (nighttime releases) for the day. Local newspapers publish these data.

c. **Atmospheric stability class:** Depending on the amount of incoming solar radiation as well as other factors, the atmosphere may be more or less turbulent at any given time. Meteorologists have defined six atmospheric stability classes, each representing a different degree of turbulence in the atmosphere. When moderate to strong incoming solar radiation heats air near the ground, causing it to rise and generating large eddies, the atmosphere is considered unstable, or relatively turbulent. Unstable conditions are associated with stability classes A and B. When solar radiation is relatively weak, air near the surface has less of a tendency to rise and less turbulence develops. In this case, the atmosphere is considered stable or less turbulent with weak winds; the stability class is E or F. Stability classes D and C represent conditions of more neutral stability, or moderate turbulence. Neutral conditions are associated with relatively strong wind speeds and moderate solar radiation. Select either A, B, C, D, E, or F by selecting the appropriate entry from the drop down list.

SURFACE WIND SPEED AT 10 METERS ABOVE GROUND		DAY			NIGHT*	
Meters per second	Miles per hour	Incoming Solar Radiation			Thinly Overcast or $\geq$ 4/8 low cloud	$\leq$ 3/8 Cloud
		Strong**	Moderate	Slight***		
< 2	<4.5	A	A-B	B		
2-3	4.5-7	A-B	B	C	E	F
3-5	7-11	B	B-C	C	D	E
5-6	11-13	C	C-D	D	D	D
>6	>13	C	D	D	D	D

    d. Precipitation present: Precipitation may take the form of hail, mist, rain, sleet, or snow. If there was precipitation present at the time of the event, check the check box in this field.

    e. Unknown weather conditions (check if a-d are all unknown): If you do not have information regarding the weather at the time of the event, select the check box available in this field. EPA recognizes that you may not have weather data for accidents that occurred in the past. You must, however, collect these data during future accident investigations.

6.9. On-site Impacts (enter numbers only): [CBI]

	Employees or contractors	Public responders	Public
6.9.a. Deaths: *			
6.9.b. Injuries: *			
6.9.c. Property damage ($): *			

6.10. Known off-site impacts (enter numbers only): [CBI]

6.10.a. Deaths: *	6.10.b. Hospitalizations: *	6.10.c. Other medical treatments: *
6.10.d. Evacuated: *	6.10.e. Sheltered-in-place: *	6.10.f. Property damage ($): *

6.10.g. Environmental damage

6.10.g.1. Fish or animal kills ☑	6.10.g.2. Tree, lawn, shrub or crop damage ☐	6.10.g.3. Water contamination ☐
6.10.g.4. Soil contamination ☐	6.10.g.5. Other (specify)	

6.9   On-site impacts:

Complete each of the following about on-site effects. Enter a number for each entry; if there were no impacts, enter 0.

   a. Deaths: Enter the number of on-site deaths that are attributable to the accident or mitigation activities. Onsite deaths include anyone (employees, contractors, responders, or others), who was killed by direct exposure to toxic concentrations, radiant heat, or overpressures from the accidental release or from indirect consequences of a vapor cloud explosion from an accidental release (e.g., flying glass debris or other projectiles). If there were no on-site deaths, enter 0. Specify the deaths as:

   i.   Employees or contractors

   ii. Public responders (example: fire department personnel)

   iii. Public (example: visitors)

   b. Injuries: An injury is any effect that results either from direct exposure to toxic concentrations, radiant heat, or overpressures from the accidental release or from indirect consequences of a vapor cloud explosion from an accidental release (e.g., a window shattering after an explosion) and that requires medical treatment or hospitalization. Medical treatment means treatment, other than first aid, administered by a physician or registered professional personnel under standing orders from a physician (OSHA OII Log, 1904.12).Your OSHA occupational injury and illness log will help complete these items for employees. Enter the appropriate value for these fields. If there were no on-site injuries, enter 0. Specify the injuries as:

   i.   Employees or contractors

   ii. Public responders (example: fire department personnel)

   iii. Public (example: visitors)

   c. Property damage ($): Estimate the value of the equipment or business structures at your facility that were damaged by the accident or mitigation activities. Record the value in US dollars. Insurance claims may provide this information. Do **not** include any losses that you may have incurred as a result of business interruption. If there was no onsite property damage or no known damage, enter 0.

6.10   Known offsite impacts:

These are impacts of which you should be aware (e.g., from media reports) or that were reported to your facility. You are **not** required to conduct an additional investigation to determine off-site impacts. Enter a number for each entry; if there were no impacts, enter 0.

   a. Deaths: Enter the number of offsite deaths that are attributable to the accident or mitigation activities. Offsite deaths include anyone who was killed by direct exposure to toxic concentrations, radiant heat, or overpressures from the accidental release or from indirect consequences of a vapor cloud explosion from an accidental release (e.g., a window shattering after an explosion). **Responders killed while on site responding to the release are considered on-site deaths and should not be reported here** (See 6.9, On-site Impacts). If there were no known offsite deaths, enter 0.

b. **Hospitalizations:** Enter the number of people requiring hospitalization. Hospitalization means any effect that result either from direct exposure to toxic concentrations, radiant heat, or overpressures from accidental releases or from indirect consequences of a vapor cloud explosion from an accidental release (e.g., a window shattering after an explosion) and that requires hospitalization (i.e., admittance to the hospital). If there were no known off-site hospitalizations, enter 0.

c. **Other medical treatments:** Enter the number of people requiring medical treatment. Medical treatment means any effect that result either from direct exposure to toxic concentrations, radiant heat, or overpressures from accidental releases or from indirect consequences of a vapor cloud explosion from an accidental release (e.g., a window shattering after an explosion) and that requires medical treatment. If there was no known medical treatment, enter 0.

d. **Evacuated:** Estimate the number of people who were evacuated to prevent exposure that might have resulted from the accident. A total count of the number of people evacuated is preferable to the number of houses evacuated. People who were ordered to move simply to improve access to the site for emergency vehicles are not considered to have been evacuated, but people who normally-occupy a building or area and who are prevented from entering or returning (i.e., blockaded) in order to prevent potentially harmful exposure should be considered to have been evacuated. If there were no evacuations, enter 0.

e. **Sheltered-in-place:** Estimate the number of people who were sheltered-in-place during the accident. Sheltering-in-place occurs when the incident commander orders residents to remain inside their home or place of work until the emergency has ended, preventing exposure to the substance. Usually these are associated with an emergency broadcast or similar method of mass notification by response agencies. If no one sheltered in place, enter 0.

f. **Property damage ($):** Estimate the value of the equipment or structures off-site that were damaged by the accident or mitigation activities. Record the value in U.S. dollars. Insurance claims may provide this information; any level of off-site property damage triggers reporting. There is no lower limit below which you would not have to report. If there was no property damage, enter 0.

g. **Environmental damage:** Indicate whether any environmental damage occurred and specify the type. The damage is not limited to environmental receptors listed in the rule. Any damage to the environment (e.g., dead or injured animals, defoliation, and water contamination) must be reported. Select all that apply.

- Fish or animal kills
- Tree, lawn, shrub, or crop damage
- Water contamination

- Soil contamination
- Other (specify)

6.11. Initiating event: * CBI

| 6.11.a. Equipment failure: ⊙ | 6.11.b. Human error: ⊙ | 6.11.c. Natural (weather conditions, earthquake): ⊙ |

6.11.d. Unknown: ⊙

6.12. Contributing factors: CBI

6.12.a. Equipment failure ☐	6.12.b. Human error ☐	6.12.c. Improper procedures ☐
6.12.d. Overpressurization ☐	6.12.e. Upset condition ☐	6.12.f. By-pass condition ☐
6.12.g. Maintenance activity/inactivity ☐	6.12.h. Process design failure ☐	6.12.i. Unsuitable equipment ☐
6.12.j. Unusual weather conditions ☐	6.12.k. Management error ☐	

6.12.l. Other (specify)

6.11   Initiating event:

Select the check box for the one initiating event that best describes the immediate cause of the accident.

a. **Equipment failure:** A device or piece of equipment failed or did not function as designed. For example, the vessel wall corroded or cracked.

b. **Human error:** An operator performed a task improperly, either by failing to take the necessary steps or by taking the wrong steps.

c. **Natural (weather conditions, earthquake):** Weather conditions, such as lightning, hail, ice storms, tornados, hurricanes, floods, high winds or earthquakes caused the accident.

d. **Unknown.**

6.12   Contributing factors:

These are factors that contributed to the accident's occurring, but were not the initiating event. If you conducted an investigation of the release, you should have identified factors that led to the initiating event. Select all that apply.

a. **Equipment failure:** A device or piece of equipment failed to contain substance or did not function as designed, thereby allowing a substance to be released.

b. **Human error:** A person performed an operation improperly or made a mistake which resulted in an accident.

c. **Improper procedures:** The procedure did not reflect the proper method of operation, the procedure omitted steps that affected the accident, or the procedure was written in a manner that allowed for misinterpretation of the instructions.

d. **Overpressurization:** The process was operated at pressures exceeding the design working pressure.

e. **Upset condition:** Release was caused by incorrect process conditions (e.g., increased temperature or pressure).

f. **By-pass condition:** The failure occurred in a pipe, channel, or valve that diverts fluid flow from the main pathway when design process or storage conditions are exceeded (e.g., overpressure). By-pass conditions may be designed to release the substance to restore acceptable process or storage conditions and prevent more severe consequences (e.g., explosion).

g. **Maintenance activity/inactivity:** The failure occurred because of maintenance activity or inactivity. An example of maintenance activity is putting the wrong gasket on a tank fitting. An example of maintenance inactivity is storage racks that remained unpainted for so long that corrosion caused the metal to fail.

h. **Process design failure:** The failure resulted from an inherent flaw in the design of the process (e.g., pressure needed to make product exceed the design pressure of the vessel).

i. **Unsuitable equipment:** The equipment used was incorrect for the process. For example, the forklift was too large for the corridors.

j. **Unusual weather conditions:** Weather conditions, such as lightning, hail, ice storms, tornados, hurricanes, floods, high winds, earthquakes and caused the accident.

k. **Management error:** The failure occurred due to any management error or management system error not included in categories a through j. Such factors may include inadequate training, inadequate oversight, inadequate hazard analysis, or other management-related factors.

l. **Other (specify).**

6.13. Off-site responders notified: *

6.13.a. Notified only: ○   6.13.b. Notified and responded: ○   6.13.c. No, not notified: ○

6.13.d. Unknown: ○   6.13.e. Other (specify):   ○

6.14. Changes introduced as a result of the accident: *

6.14.a. Improved/upgraded equipment ☐   6.14.b. Revised maintenance ☐   6.14.c. Revised training ☐

6.14.d. Revised operating procedures ☐   6.14.e. New process controls ☐   6.14.f. New mitigation systems ☐

6.14.g. Revised emergency response plan ☐   6.14.h. Changed process ☐   6.14.i. Reduced inventory ☐

6.14.j. None ☐   6.14.k. Other (specify)

[Discard Changes]   [Delete Accident]   [Save and Return]

6.13   Off-site responders notified:

Indicate whether response agencies (e.g., police, fire, medical services) were notified. Check one of the following boxes:

a. Notified only
b. Notified and responded
c. No, not notified
d. Unknown
e. Other (specify)

6.14   Changes introduced as a result of the accident:

Indicate any measures that you have taken at the facility to prevent recurrence of the accident. Select at least one.

a. **Improved/upgraded equipment:** A device or piece of equipment that did not function as designed was repaired or replaced.
b. **Revised maintenance:** Maintenance procedures were clarified or changed to ensure appropriate and timely maintenance including inspection and testing (i.e., increasing the frequency of inspection or adding a testing method).
c. **Revised training:** Training programs were clarified or changed to ensure that employees and contract employees are aware of and are practicing correct safety and administrative procedures.
d. **Revised operating procedures:** Operating procedures were clarified or changed to ensure that employees and contract employees are trained on appropriate operating procedures.

e. **New process controls:** New process designs and controls were installed to correct problems and prevent recurrence of an accidental release.

f. **New mitigation systems:** New mitigation systems were initiated to limit the severity of accidental releases.

g. **Revised emergency response plan:** The emergency response plan was revised.

h. **Changed process:** Process was altered to reduce the risk (e.g., process chemistry was changed).

i. **Reduced inventory:** Inventory was reduced at the facility to reduce the potential release quantities and the magnitude of the hazard.

j. **None:** No changes initiated at facility as a result of the accident (i.e., none were necessary or technically feasible). There may be some accidents that could not have been prevented because they were caused by events that are too rare to merit additional steps. For example, if a tornado hit your facility and you are located in an area where tornados are very rare, it may not be reasonable to design a "tornado-proof" process even if it is technically feasible.

k. **Other (specify).**

Section 7: Prevention Program: Program Level 3

Complete this section for each prevention program you report for a Program 3 process. You will only be able to add or update a scenario if a Program Level 3 process is present within *Section 1. Processes* section.

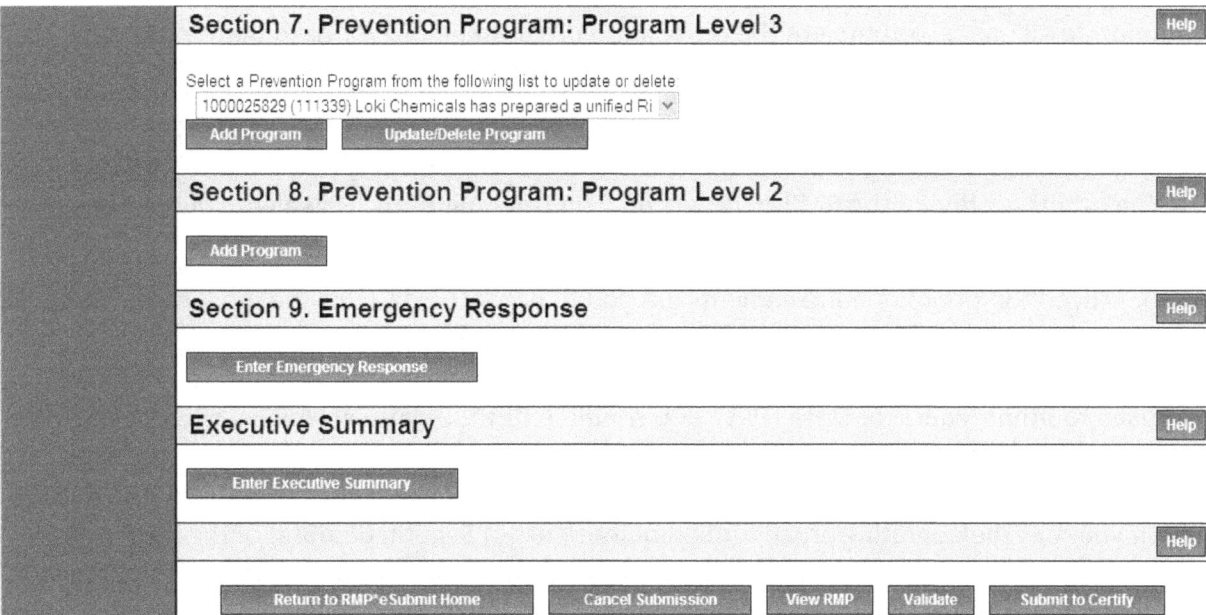

How Must Prevention Program Data Be Reported?

Prevention program data must be reported on a process-by-process basis. In other words, you must fill out the prevention program section of the RMP for each Program 2 or Program 3 process you have that is subject to the RMP rule.

How to report the prevention program for a process depends on how many units the process contains and whether the prevention program applies different safeguards to different units in the process. The RMP rule broadly defines "process" to include interconnected or co-located production and storage units. Under the definition, multiple units and, in some cases, whole sources may be a single "process" for purposes of the RMP rule. For multiple unit processes, EPA recognizes that prevention program implementation may involve different safeguards for different units in the process. For example, different production units may have different operating procedures. At the same time, some safeguards, such as management of change procedures, may apply to all the units in the process.

If your process consists of two or more units and different safeguards apply to different units in that process, you can report the prevention program for that process in one of the two following ways. You must, however, use one of the two ways to report your program.

Use the description field in the prevention program to describe in narrative form how your prevention program is implemented with respect to the different units in the process. You could start by listing the common prevention program elements you implement for all of the units (e.g., use of an alarm system or standard management of change procedures). You would then indicate what additional prevention program elements you employ for specified units (e.g., use of a dike for certain process units).

Report your prevention program for the process on a unit-by-unit basis by filling out the prevention program portion of the RMP for every unit in the process, rather than for the process as a whole. That way, the differences in the program as it relates to each unit will be clear from the report. However, as noted above, some aspects of a prevention program may be common to all units. To complete the prevention program record for each unit, enter the remaining data which is unique to each.

If your process consists of only one unit, or you apply every element of your prevention program to all the units in the process, you are not required to complete the description section of this portion of the RMP or report on a unit-by-unit basis. However, you may use the description field to elaborate on your prevention program.

Many prevention program data elements ask you to enter the date for the most recent "review or revision" of a prevention program element required by Part 68. For your first RMP submission, if you are subject to prevention program requirements only under the RMP rule (as opposed to other federal or state laws), you should enter the date by which you completed the prevention program element being addressed. For instance, for data element number 7.5, "Date of most recent review or revision of operating procedures," you should enter the date by which you met the operating procedures requirements of Section 68.69(a) of the RMP rule (if applicable to you). Since this requirement must be met by the time your first RMP is due, you may enter the date you complete or submit your first RMP. In the case of data element number 7.3 ("Date on which safety information was last reviewed or revised"), you should enter the date you met the requirement of Section 68.65(a) (if applicable to you), since Section 68.65(a) requires you to meet the requirement before you conduct the process hazard analysis for the process.

If you are subject to prevention program requirements under other federal or state laws, you may be in compliance with RMP prevention program requirements as a result of complying with the other laws. Sources subject to OSHA PSM, for example, may already meet RMP prevention program requirements for Program 3 processes, since those requirements are nearly identical to OSHA PSM prevention program requirements. For your first RMP submission, if you have fulfilled RMP prevention program requirements in complying with other federal or state laws, you should enter the date you complied with the requirement or the date you last reviewed or revised the relevant aspect of your program, whichever is later. For example, OSHA PSM and the RMP rule both require covered sources to compile and update (under specified circumstances) process safety information. If you previously compiled the information for purposes of complying with OSHA PSM and you have not updated it since, you should enter the date you compiled it for OSHA in your RMP. If you have updated the information since compiling it, you should enter the date of the update.

For subsequent RMP submissions, you should enter the date by which you completed any review or revision of a prevention program element. Several prevention program elements must be reviewed and, if necessary, revised following a change affecting the process (see, e.g., requirement to update safety information in Section 68.75(d)). Under the compliance audit requirement of Sections 68.58 or 68.79 of the RMP Program, all prevention program elements must be reviewed and, if appropriate, revised every three years. When you re-submit your next RMP (due every 5 years or sooner based on the requirements in Section 68.190), you are required to fully update and certify all nine sections of the RMP. If, by the time you re-submit, you have reviewed or revised one or more prevention program elements as a result of a change or an audit, you must enter the date of your review or revision.

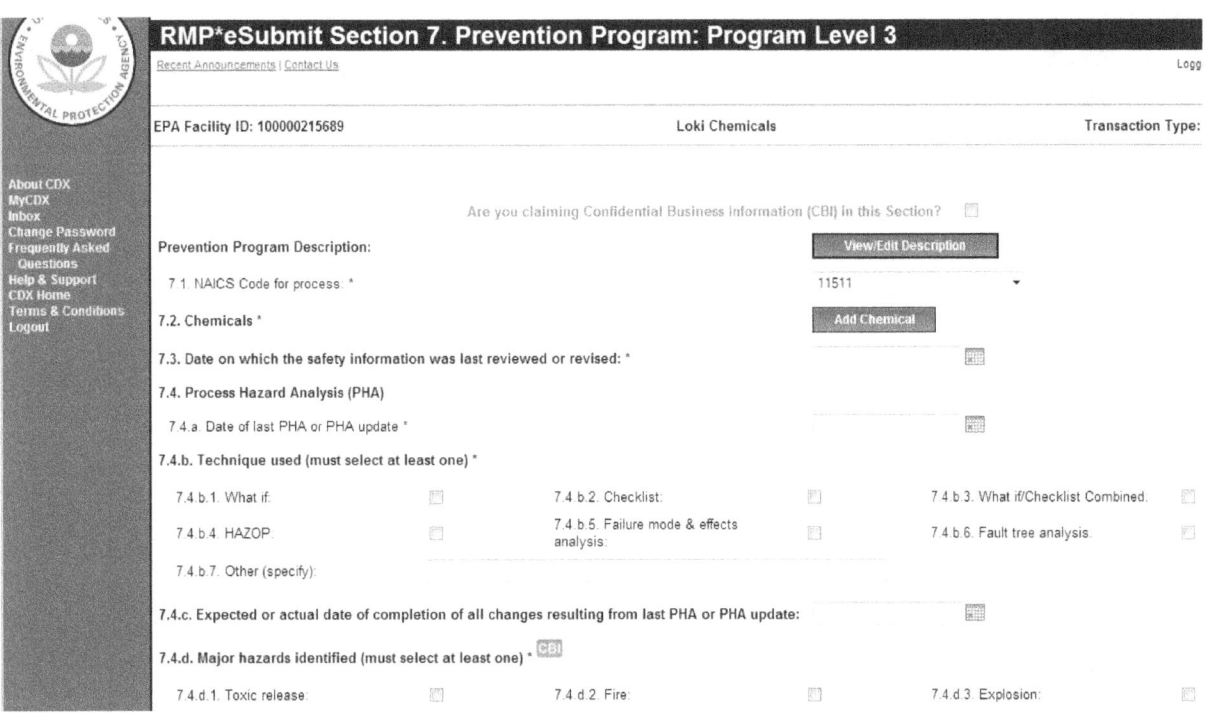

Section 7: Prevention Program: Program Level 3

The following is a discussion of each element in *Section 7. Prevention Program: Program Level 3.*

Prevention Program Description:

If different safeguards apply to different units in your process, use this field to explain how the prevention program for the process relates to the different units in the process. For example:

> *"This process includes three interconnected production units, A, B, and C. Everything in this prevention program applies to all three units, with the following exceptions:*

- The dates of the PHA, which are 01/02/08, 6/5/07 and 4/3/07 for units A, B, and C, respectively.
- Production unit A uses only a scrubber as a process control, while units B and C have relief valves and scrubbers.
- The water curtain indicated as a mitigation measure applies only to production unit C."

If you have so many "exceptions" that it gets too complicated to explain as above, but you still have many common data elements, you can report your prevention program on a unit-by-unit basis. To complete the prevention program record for each unit, provide the remaining data which is unique to each.

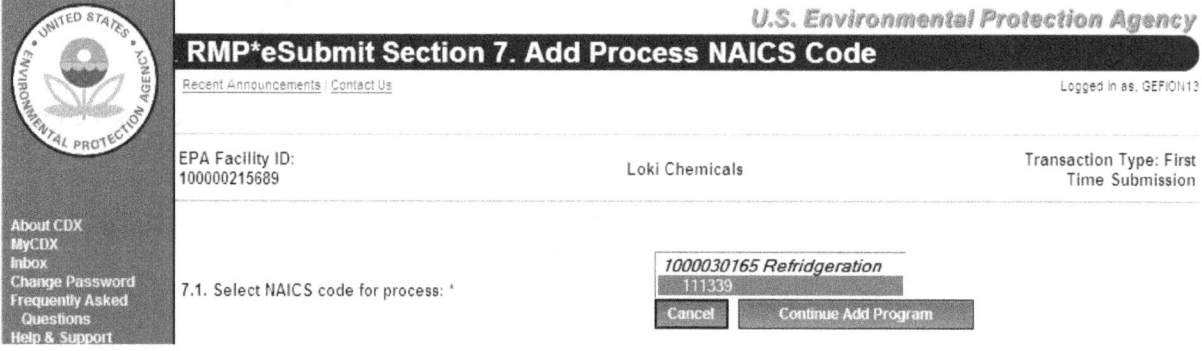

7.1   NAICS code for the process:

Provide the NAICS code that most closely corresponds to the process; it will not necessarily be the same NAICS code as your facility as a whole. The NAICS code that you choose must be one that you've already entered in the Registration Section for the covered process. RMP*eSubmit contains a list of the NAICS codes that you have already entered for your registered processes as a pick list for this data element.

7.2   Chemicals:

For each prevention program, provide the names of all regulated substances held above the threshold in the covered units. If you have an NFPA-4 flammable mixture containing regulated flammables, you may list it as a "flammable mixture." You do not need to list the individual substances in the flammable mixture.

7.3   Date on which the safety information was last reviewed or revised:

The safety information requirements for Program 3 processes can be found at 40 CFR68.65. For your first RMP, provide the date by which you complied with the safety information requirements of Section 68.65(a) (compile safety information) for the process. For subsequent RMPs, provide the date the safety information was most recently reviewed or revised. Safety information may be reviewed or revised as a result of, among other things, a change to the process (see Section 68.75(d)) or a periodic audit of the prevention program (see Section 68.79(a)). If the safety information was not reviewed or revised since the previous RMP was submitted, re-enter the date provided in the previous Risk Management Plan (RMP).

7.4   Process Hazard Analysis (PHA):

a. Date of last PHA or PHA update:

Provide the date you completed or updated your most recent PHA.

b. Technique used (must select at least one): Check any of the following techniques that you used to evaluate the hazards of your process or part of the process (see Chapter 8, Appendix A, of the *General Guidance for Risk Management Programs* for a description of these techniques). Select at least one by clicking on the check box for each corresponding technique:

- What If
- Checklist
- What If/Checklist Combined
- HAZOP
- Failure Mode & Effects Analysis
- Fault Tree Analysis
- Other (specify)

c. Expected or actual date of completion of all changes resulting from last PHA or PHA update: Provide the date you completed or expect to complete any changes resulting from the PHA. This may be blank if there were no changes.

Major hazards identified: Select any major hazards that were identified for the Program 3 process or part of the process as a result of the PHA. Major hazards are described below. Select at least one by clicking on the check box for each corresponding technique:

7.4.D. MAJOR HAZARD	DESCRIPTION
1. Toxic release	If an accidental release occurred, a regulated toxic substance could be released. For listed toxics, a toxic release will always be a major hazard.
2. Fire process	Upsets, leaks, equipment failure, etc., could result in a fire. For listed flammables, fire will always be a major hazard. Fire may also be a hazard for some listed toxics and in some processes could cause a toxic release.
3. Explosion	Confined or unconfined vapor cloud explosions. For listed flammables, explosion will generally be a major hazard. Explosion may also be a hazard for toxics, especially those handled under extreme conditions.
4. Runaway reaction	An uncontrolled reaction that proceeds at an increasing rate.
5. Polymerization	A chemical reaction that produces the bonding of two or more monomers.
6. Overpressurization	Instantaneous energy release or detonation.
7. Corrosion	Corrosion could lead to destruction of equipment and release of a regulated substance. Corrosion is likely to be a major hazard for substances identified as corrosives on MSDSs unless the equipment used limits the hazard.
8. Overfilling	Filling a tank or vessel beyond its maximum safe capacity.
9. Contamination	A release could occur if inappropriate substances are introduced into storage or process vessels. Contamination may be a major hazard when controlling inappropriate substances (e.g., H_2O) is difficult.
10. Equipment failure	Equipment failure is likely to be a major hazard for most processes, because such failure could lead to a release. Equipment failure includes cracks, weld failures, disk failures, ruptures, pump/gauge/control system failures, etc.
11. Loss of cooling, heating, electricity, instrument air	These losses could be major hazards, if they could lead to releases. For example, loss of cooling could lead to an increase in pressure and failure of a vessel or pipe and a loss of heating or power could lead to unstable processes. These conditions are less likely to be major hazards for substances handled at atmospheric temperatures and pressures.
12. Earthquake	Report earthquakes as a major hazard only if they occur or are likely to occur at your site such that you plan and design for them.
13. Floods (Flood Plain)	Report floods as a major hazard only if they occur or are likely to occur at your site such that you plan and design for them.
14. Tornadoes	Report tornadoes as a major hazard only if they occur or are likely to occur at your site such that you plan and design for them.
15. Hurricanes	Report hurricanes as a major hazard only if they occur or are likely to occur at your site such that you plan and design for them.
16. Other	Specify any other major hazards not listed above.

d. Process controls in use: Select all of the process controls used on the process or part of the process. Process controls are equipment and associated procedures used to prevent or limit releases and are described below. If none are applicable, check *None*.

7.4.E. PROCESS CONTROL	DESCRIPTION
1. Vents	An opening provided for the discharge of pressure or release of pressure from tanks, vessels, or processing equipment.
2. Relief Valves	A valve that relieves pressure beyond a specified limit and recloses to normal operating pressure upon return.
3. Check Valves	A device for automatically limiting the flow in a piping system to a single direction.
4. Scrubbers	A pre-release protection measure that uses water or aqueous mixtures containing scrubbing reagents to remove discharging liquids and may treat the discharging chemical.
5. Flares	A pre-release protection measure used for flammable gases and vapors to remove and possibly treat discharged liquids.
6. Manual Shutoffs	Manual controls of the shutoff flow to a pipe or vessel.
7. Automatic Shutoffs	Controls the shutoff flow to a pipe or vessel and are triggered automatically when process conditions are exceeded.
8. Interlocks	A switch or other device that prevents activation of a piece of equipment when a protective door is open or some other hazard exists.
9. Alarms and Procedures	Systems that trigger a warning device after the occurrence of a hazardous condition and procedures to activate an alarm system.
10. Keyed Bypass	A bypass system that is activated by a control signal.
11. Emergency Air Supply	A backup system to provide air to a process when the regular air supply fails.
12. Emergency Power	Backup power systems.
13. Backup Pump	A secondary pump intended to serve the same function as the primary pump if the primary pump fails.
14. Grounding Equipment and Bonding	Devices that ground and bond electrical equipment to avoid explosions and to provide a good electrical path to the ground.
15. Inhibitor Addition	A substance that is added to a reaction that is capable of stopping or retarding a chemical reaction.
16. Rupture Disks	A device that relieves pressure beyond a specified limit.
17. Excess Flow Device	Flow-limiting equipment that protects downstream equipment from surges.
18. Quench System	A system that cools by removing excess heat or immersing liquid into a cooling medium.

7.4.E. PROCESS CONTROL	DESCRIPTION
19. Purge System	A system that replaces the atmosphere in a container with an inert substance to prevent the formations of an explosive mixture.
20. None	None are applicable.
21. Other	Specify any other process controls that you may use on your process and that are not specified above.

e. Mitigation systems in use: Select all of the mitigation systems you have in place to control a release from the process or part of the process. Mitigation systems are described below. If none are applicable, check *None*.

7.4.F. MITIGATION SYSTEMS	DESCRIPTION
1. Sprinkler systems	A system for protecting a building against a fire by means of overhead pipes that release an extinguishing material through heat activated outlets.
2. Dikes	Upsets, leaks, equipment failure, etc., could result in a fire. For listed flammables, fire will always be a major hazard. Fire may also be a hazard for some listed toxics and in some processes could cause a toxic release.
3. Fire walls	A wall constructed to prevent the spread of fire.
4. Blast walls	A heavy wall used to isolate buildings or areas that contain highly combustible or explosive materials.
5. Deluge system	A system to overflow an area with a release of water or other extinguishing fluid.
6. Water curtain	A spray of water from a horizontal pipe through nozzles. The curtain may be activated manually or automatically.
7. Enclosure	Something that facilitates the physical containment of a release within a structure (e.g., a building).
8. Neutralization	Controlling a release by neutralizing the released chemical.
9. None	None are applicable.
10. Other	Specify any other mitigation systems you may have in place on your process and that are not listed above.

f. Monitoring/detection systems in use: Select all of the monitoring and detection systems you have installed to detect a release of a regulated substance from the process or part of the process. Monitoring and detection systems are described below. If none are applicable, check *None*.

7.4.G. MONITORING & DETECTION SYSTEMS	DESCRIPTION
1. Process area detectors	Detection systems located on or close to process equipment. Detection systems include indicator tubes, and chromatographic, spectrometric, electrochemical, and colorimetric gas analysis.
2. Perimeter Monitors	Integrated detection networks at the source boundary. Detection systems can include fluorescent SO_2 analyzers, photoelectric tape sensors, or electrolytic chlorine detectors.
3. None	None are applicable.
4. Other	Specify any other monitoring and detection systems you have in place and that are not listed above.

g. Changes since last PHA or PHA update: Select all of the changes made to the process or part of the process since the last PHA. If none are applicable, check *None*.

7.4.H. CHANGES SINCE LAST PHA / PHA UPDATE	DESCRIPTION
1. Reduction in chemical inventory	A decrease in the quantity of regulated substances stored on-site.
2. Increase in chemical inventory	An increase in the quantity of regulated substances stored on-site.
3. Change in process parameters	Examples of changes in process parameters include an increase or decrease in temperature, pressure, flow rates, etc.
4. Installation of process controls	The addition of controls such as those described in 7.4 e. or 8.4 d.
5. Installation of process detection systems	The addition of systems such as those described in 7.4 g. or 8.4 f.
6. Installation of perimeter monitoring systems	The addition of systems such as those described in 7.4 g. or 8.4 f.
7. Installation of mitigation systems	The addition of systems such as those described in 7.4 f. or 8.4 e.
8. None recommended	Select *None recommended* if the PHA team did not recommend any changes to the process.
9. None	None are applicable.
10. Other (specify)	Specify any other changes made to the process since the last PHA that are not listed above.

7.5. Date of most recent review or revision of operating procedures: *

7.6. Training

   7.6.a. Date of most recent review or revision of training programs: *

   7.6.b. Type of training provided (must select at least one) *

      7.6.b.1. Classroom: ☐   7.6.b.2. On the job: ☑

      7.6.b.3. Other (specify):

   7.6.c. Type of competency testing used (must select at least one) *

      7.6.c.1. Written test: ☐   7.6.c.2. Oral test: ☐   7.6.c.3. Demonstration:

      7.6.c.4. Observation: ☑   7.6.c.5. Other (specify):

7.7. Maintenance

   7.7.a. Date of most recent review or revision of maintenance procedures: *

   7.7.b. Date of most recent equipment inspection or test: *

   7.7.c. Equipment most recently inspected or tested (equipment list): *

7.8. Management of change

   7.8.a. Date of most recent changes that triggered management of change procedures:

   7.8.b. Date of most recent review or revision of management of change procedures: *

7.5 *Date of most recent review or revision of operating procedures:*

The operating procedures requirements for Program 3 processes can be found at 40 CFR 68.69. For your first RMP, provide the date by which you complied with the requirements of Section 68.69(a) (develop and implement written procedures) for the process. For subsequent RMPs, provide the date of the most recent review or revision of the operating procedures. Operating procedures may be reviewed or revised as a result of, among other things, a change to the process (see Sections 68.69(c) and 68.75(e)), annual certification of the operating procedures (see Section 68.69(c)), or a periodic audit of the prevention program (see Section 68.79(a)).

7.6 *Training:*

   a. **Date of most recent review or revision of training programs:** For your first RMP, provide the date by which you ensured that the training you provide the employees operating the process meets the requirements of Section 68.71(a). For subsequent RMPs, provide the date of the most recent review or revision of the training you provide. Training programs may be reviewed or revised as a result of, among other things, a change to the process (see Section 68.75(c)) or a periodic audit of the prevention program (see Section 68.79(a)). If the training was not reviewed or revised since the previous RMP was submitted, re-enter the date provided in the previous RMP.

   b. **Type of training provided:** Select the type of training provided (select all that apply). Training information can be found in the RMP regulation at 40 CFR 68.54 and 68.71.

   c. **Type of competency testing used:** Indicate the type of competency test used: written test, oral test, demonstration, or observation by selecting the appropriate check box(es). Competency tests are used to determine and evaluate comprehension of the training materials. Training information can be found in the RMP regulation at 40 CFR 68.54 and 68.71.

7.7   *Maintenance:*

a. **The date that you most recently reviewed or revised the maintenance procedures:** For your first RMP, provide the date by which you complied with the requirements of Section 68.73(b) (establish and implement written maintenance procedures) for the process. For subsequent RMPs, provide the date of the most recent review or revision of the maintenance procedures. Maintenance procedures may be reviewed or revised as a result of, among other things, a change to the process (see Section 68.75(c)) or a periodic audit of the prevention program (see Section 68.79(a)). If the procedures were not reviewed or revised since the previous RMP was submitted, re-enter the date provided in the previous RMP.

b. **The date of the most recent equipment inspection or test:** Provide the appropriate date. Maintenance information can be found in the RMP regulation at 40 CFR 68.56 and 68.73.

c. **The equipment that was inspected or tested (list equipment):** Specify the equipment inspected or tested. Maintenance information can be found in the RMP regulation at 40 CFR 68.56 and 68.73.

7.8   *Management of change:*

a. **The date of the most recent change (if any) that triggered the management of change procedure:** Provide the appropriate date. Management of Change information can be found in the RMP regulation at 40 CFR 68.75.

b. **The date that you most recently reviewed or revised the management of change procedures at your site:** For your first RMP, provide the date by which you complied with the requirements of Section 68.75(a) (establish and implement written procedures) for the process. For subsequent RMPs, provide the date of the most recent review or revision of the procedures. Management of change procedures may be reviewed or revised as a result of, among other things, a periodic audit of the prevention program (see Section 68.79(a)). If the procedures were not reviewed or revised since the previous RMP was submitted, retain the date provided in the previous RMP.

7.9. Date of most recent pre-startup review:

7.10. Compliance audits

   7.10.a. Date of most recent compliance audits:

   7.10.b. Expected or actual date of completion of all changes resulting from the most recent compliance audits:

7.11. Incident investigation

   7.11.a. Date of most recent incident investigation (if any):

   7.11.b. Expected or actual date of completion of all changes resulting from the incident investigation:

7.12. Date of most recent review or revision of employee participation plans: *

7.13. Date of most recent review or revision of hot work permit procedures: *

7.14. Date of most recent review or revision of contractor safety procedures:

7.15. Date of most recent evaluation of contractor safety performance:

| Discard Changes | Delete Program | Save and Return |

7.9   Date of most recent pre-startup review:

The pre-startup review requirements for Program 3 processes can be found at 40 CFR 68.77. Provide the date of the most recent pre-startup review (if any) for this process.

7.10   Compliance audits:

a. Date of most recent compliance audit: Provide the date of your most recent compliance audit. If you have not conducted a compliance audit prior to your first submission, leave these fields blank.

NOTE: A compliance audit is required every three years. Compliance audit information can be found in the RMP regulation at 40 CFR 68.58 and 68.79.

b. Expected or actual date of completion of all changes resulting from the compliance audit: This may be left blank if there were no changes. Incident Investigation information can be found in the RMP regulation at 40 CFR 68.60 and 68.81.

7.11   Incident investigation:

The incident investigation requirements for Program 3 processes can be found at 40 CFR 68.60 and 68.81.

a. Date of your most recent incident investigation (if any): Provide the date of your most recent incident investigation, if any. If you have not had an incident investigation, leave this field blank.

b. The expected or actual date of completion of all changes resulting from the incident investigation: Provide the expected or actual date of completion of all changes resulting from the incident investigation. This may be left blank if there were no changes or all changes are complete.

7.12 Date of most recent review or revision of employee participation plans:

The employee participation requirements for Program 3 processes can be found at 40 CFR 68.83. For your first RMP, provide the date by which you complied with the requirements of Section 68.83(a) (develop a written plan) for the process. For subsequent RMPs, provide the date of the most recent review or revision of the plan. Employee participation plans may be reviewed or revised as a result of, among other things, a periodic audit of the prevention program (see Section 68.79(a)). If the plan was not reviewed or revised since the previous RMP was submitted, retain the date provided in the previous RMP.

7.13 Date of most recent review or revision of hot work permit procedures:

The hot work permit requirements for Program 3 processes can be found at 40 CFR 68.85. For your first RMP, provide the date by which you ensured that you comply with the requirements of Section 68.85. For subsequent RMPs, provide the date of the most recent review or revision of your permit procedures. Hot work permit procedures may be reviewed or revised as a result of, among other things, a periodic audit of the prevention program (see Section 68.79(a)). If the procedures were not reviewed or revised since the previous RMP was submitted, retain the date provided in the previous RMP.

7.14 Date of most recent review or revision of contractor safety procedures:

The contractor safety requirements for Program 3 processes can be found at 40 CFR 68.87. Leave this field blank if you do not have any contractors. Otherwise, for your first RMP, provide the date by which you complied with the requirements of Section 68.87(b) (4) (develop and implement safe work practices for contractors) for the process. For subsequent RMPs, provide the date of the most recent review or revision of the procedures. Contractor safety procedures may be reviewed or revised as a result of, among other things, a change to the process (see Section 68.75(c)) or a periodic audit of the prevention program (see Section 68.79(a)). If the procedures were not reviewed or revised since the previous RMP was submitted, retain the date provided in the previous RMP.

7.15 Date of most recent evaluation of contractor safety performance:

Leave this field blank if you do not have any contractors or have not yet evaluated contractor performance. Otherwise, provide the date of your most recent evaluation of contractor safety performance. If you have more than one contractor involved in operating or maintaining the process, provide the date that you completed your evaluations of all the contractors.

Section 8: Prevention Program: Program Level 2

Complete this section for each prevention program you report for a Program 2 process. You will only be able to add or update a scenario if a Program Level 2 process is present within *Section 1. Processes* section.

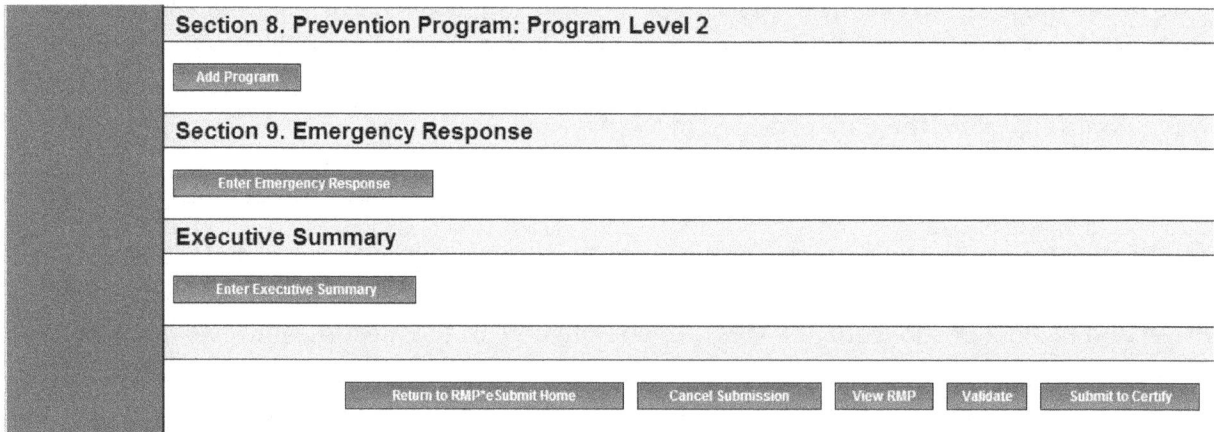

How Must Prevention Program Data Be Reported?

Prevention program data must be reported on a process-by-process basis. In other words, you must fill out the prevention program section of the RMP for each Program 2 or Program 3 process you have that is subject to the RMP rule.

How to report the prevention program for a process depends on how many units the process contains and whether the prevention program applies different safeguards to different units in the process. The RMP rule broadly defines "process" to include interconnected or co-located production and storage units. Under the definition, multiple units and, in some cases, whole sources may be a single "process" for purposes of the RMP rule. For multiple unit processes, EPA recognizes that prevention program implementation may involve different safeguards for different units in the process. For example, different production units may have different operating procedures. At the same time, some safeguards, such as management of change procedures, may apply to all the units in the process.

If your process consists of two or more units and different safeguards apply to different units in that process, you can report the prevention program for that process in one of the two following ways. You must, however, use one of the two ways to report your program.

Use the description field in the prevention program to describe in narrative form how your prevention program is implemented with respect to the different units in the process. You could start by listing the common prevention program elements you implement for all of the units (e.g., use of an alarm system or standard management of change procedures). You would then indicate what additional prevention program elements you employ for specified units (e.g., use of a dike for certain process units).

Report your prevention program for the process on a unit-by-unit basis by filling out the prevention program portion of the RMP for every unit in the process, rather than for the process as a whole. That way, the differences in the program as it relates to each unit will be clear from the report. However, as noted above, some aspects of a prevention program may be common to all units. To complete the prevention program record for each unit, enter the remaining data which is unique to each.

If your process consists of only one unit, or you apply every element of your prevention program to all the units in the process, you are not required to complete the description section of this portion of the RMP or report on a unit-by-unit basis. However, you may use the description field to elaborate on your prevention program.

Many prevention program data elements ask you to enter the date for the most recent "review or revision" of a prevention program element required by Part 68. For your first RMP submission, if you are subject to prevention program requirements only under the RMP rule (as opposed to other federal or state laws), you should enter the date by which you completed the prevention program element being addressed. For instance, for data element number 7.5, "Date of most recent review or revision of operating procedures," you should enter the date by which you met the operating procedures requirements of Section 68.69(a) of the RMP rule (if applicable to you). Since this requirement must be met by the time your first RMP is due, you may enter the date you complete or submit your first RMP. In the case of data element number 7.3 ("Date on which safety information was last reviewed or revised"), you should enter the date you met the requirement of Section 68.65(a) (if applicable to you), since Section 68.65(a) requires you to meet the requirement before you conduct the process hazard analysis for the process.

If you are subject to prevention program requirements under other federal or state laws, you may be in compliance with RMP prevention program requirements as a result of complying with the other laws. Sources subject to OSHA PSM, for example, may already meet RMP prevention program requirements for Program 3 processes, since those requirements are nearly identical to OSHA PSM prevention program requirements. For your first RMP submission, if you have fulfilled RMP prevention program requirements in complying with other federal or state laws, you should enter the date you complied with the requirement or the date you last reviewed or revised the relevant aspect of your program, whichever is later. For example, OSHA PSM and the RMP rule both require covered sources to compile and update (under specified circumstances) process safety information. If you previously compiled the information for purposes of complying with OSHA PSM and you have not updated it since, you should enter the date you compiled it for OSHA in your RMP. If you have updated the information since compiling it, you should enter the date of the update.

For subsequent RMP submissions, you should enter the date by which you completed any review or revision of a prevention program element. Several prevention program elements must be reviewed and, if necessary, revised following a change affecting the process (see, e.g., requirement to update safety information in Section 68.75(d)). Under the compliance audit requirement of Sections 68.58 or 68.79 of the RMP Program, all prevention program elements must be reviewed and, if appropriate, revised every three years. When you re-submit your next RMP (due every 5 years or sooner based on the requirements in Section 68.190), you are required to fully update and certify all nine sections of the RMP. If, by the time you re-submit, you have reviewed or revised one or more prevention program elements as a result of a change or an audit, you must enter the date of your review or revision.

Complete this section for each prevention program you report for a Program 2 process.

Prevention Program Description:

If different safeguards apply to different units in your process, use this field to explain how the prevention program for the process relates to the different units in the process. For example:

> *"This process includes three interconnected production units, A, B, and C. Everything in this prevention program applies to all three units, with the following exceptions:*

- The dates of the PHA, which are 01/02/08, 6/5/07 and 4/3/07 for units A, B and C, respectively.
- Production unit A uses only a scrubber as a process control, while units B and C have relief valves and scrubbers.
- The water curtain indicated as a mitigation measure applies only to production unit C."

If you have so many "exceptions" that it gets too complicated to explain as above, but you still have many common data elements, you can report your prevention program on a unit-by-unit basis. To complete the prevention program record for each unit, provide the remaining data which is unique to each.

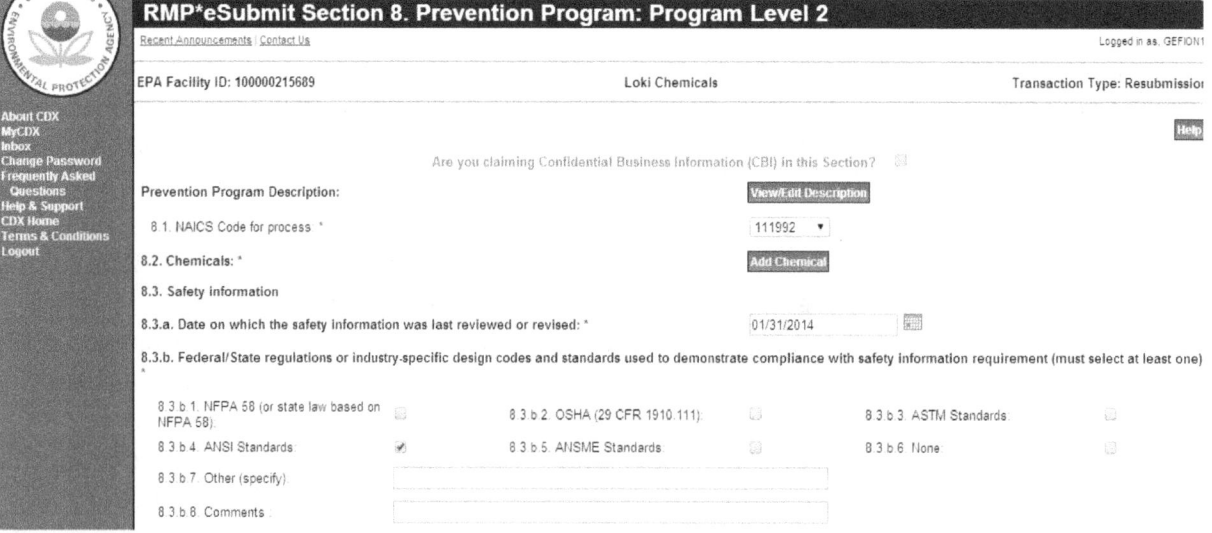

8.1   NAICS code for the process:

Provide the NAICS code that most closely corresponds to the process; it will not necessarily be the same NAICS code as your facility as a whole. The NAICS code that you choose must be one that you've already entered in the *Section 1. Registration* for the covered process. RMP*eSubmit contains a list of the NAICS codes that you have already entered for your registered processes as a list for this data element.

8.2   Chemicals:

For each prevention program, provide the names of all regulated substances held above the threshold in the covered units. If you have an NFPA-4 flammable mixture containing regulated flammables, you may list it as a "flammable mixture." You do not need to list the individual substances in the flammable mixture.

8.3   Safety information:

a.   The date that you most recently reviewed or revised the safety information:

For your first RMP, provide the date by which you complied with the safety information requirements of 40 CFR 68.48(a) (compile safety information) for the process. For subsequent RMPs, provide the date the information was most recently reviewed or revised. Safety information may be reviewed or revised as a result of, among other things, a major change to the process (see Section 68.48(c)) or a periodic audit of the prevention program (see Section 68.58(a)). If the safety information was not reviewed or revised since the previous RMP was submitted, retain the date provided in the previous RMP.

b.   Federal or state regulations or industry-specific design codes and standards used to demonstrate compliance with the safety information requirement (select at least one):

Are you subject to any of the following federal or state regulations? Do you use any of the following industry-specific design codes and standards to demonstrate compliance with the safety information requirement? If none are applicable, check *None*.

b.1.   NFPA 58 (or state law based on NFPA 58):

NFPA stands for National Fire Protection Association; NFPA 58 is a propane (LP gas) handling code. Note that state propane laws are generally based on NFPA 58. Select NFPA 58 if your process is subject to a state or local law based on NFPA 58 or if you follow NFPA 58 in any event.

b.2.   OSHA (29 CFR 1910.111):

OSHA's rule for operations handling anhydrous ammonia, select 29 CFR 1910.111 if your process is subject to this rule.

b.3.   ASTM Standards:

Select this if you follow American Society of Testing Materials standards. ASTM establishes standards for materials, products, systems, test methods, specifications, classifications, definitions, and recommended practices.

b.4.   ANSI Standards:

Select this if you follow American National Standards Institute standards. ANSI nationally coordinates voluntary standards. ANSI standards cover areas such as definitions, terminology, symbols, and abbreviations; materials, performance characteristics, procedure, and methods of rating; methods of testing and analysis; size, weight, and volume, safety, health, and building construction.

b.5.   ASME Standards:

Select this if you follow American Society of Mechanical Engineers standards. ASME conducts research and develops boiler, pressure vessel, and power test codes. It also develops safety codes and standards for equipment.

b.6.   None:

If your facility does not apply to the Program 2 process any national standards such as those noted above, and is not subject to any federal or state rules or laws such as those noted above, select *None*.

b.7.   Other (specify):

If you apply any other standards to your process safety equipment, select Other and specify the standards you apply. Some examples of other standards include the National Electrical Manufacturers Association (NEMA) standards and the American Petroleum Institute (API) standards. There may also be other codes that apply.

b.8.   Comments:

In this section, please explain how federal, state, or local regulations or industry-specific design codes and standards are being used to demonstrate compliance with the safety information requirement.

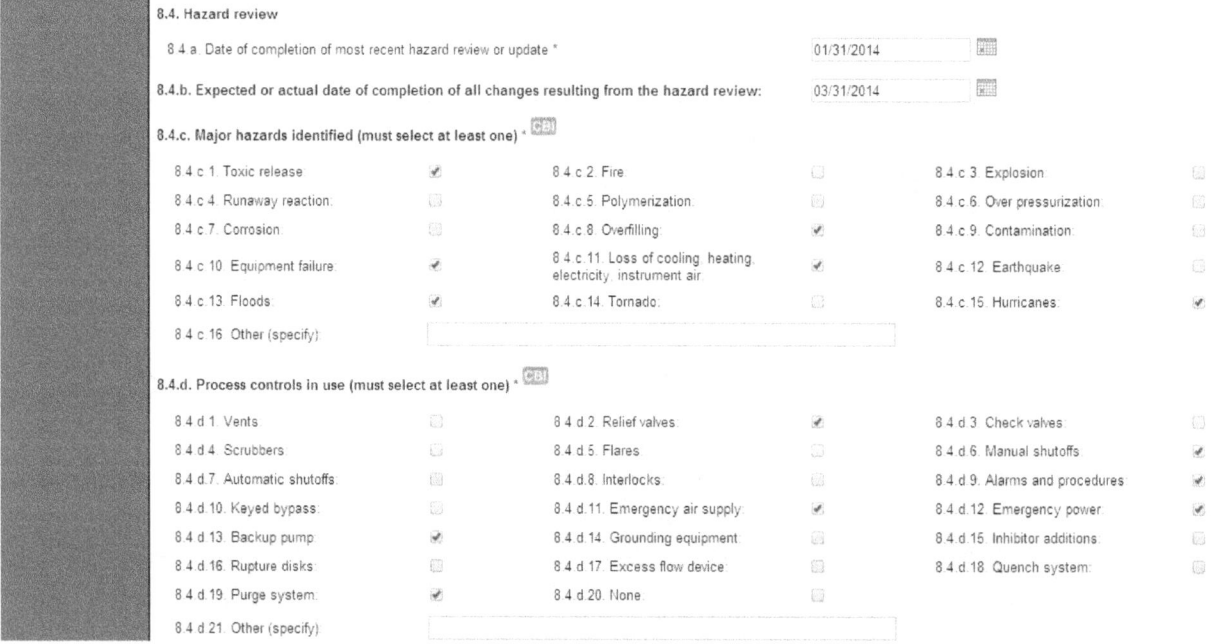

8.4   Hazard review:

The hazard review requirements for Program 2 processes can be found at 40 CFR 68.50.

a. The date of completion of the most recent hazard review or update:

Provide the date of completion of the most recent hazard review or update (must be within the five years prior to submission of the RMP).

b. The expected or actual date of completion of all changes resulting from the hazard review:

This may be left blank if there were no changes.

c. Major hazards identified:

Provide all major hazards that were identified for the Program 2 process or part of the process at your facility as a result of the hazard review. Major hazards are described in 7.4 d.

d. Process controls in use:

Provide all process controls used on this Program 2 process or part of the process. Process controls are equipment and associated procedures used to prevent or limit releases. If none are applicable, check *None*. Process controls are described in 7.4 e.

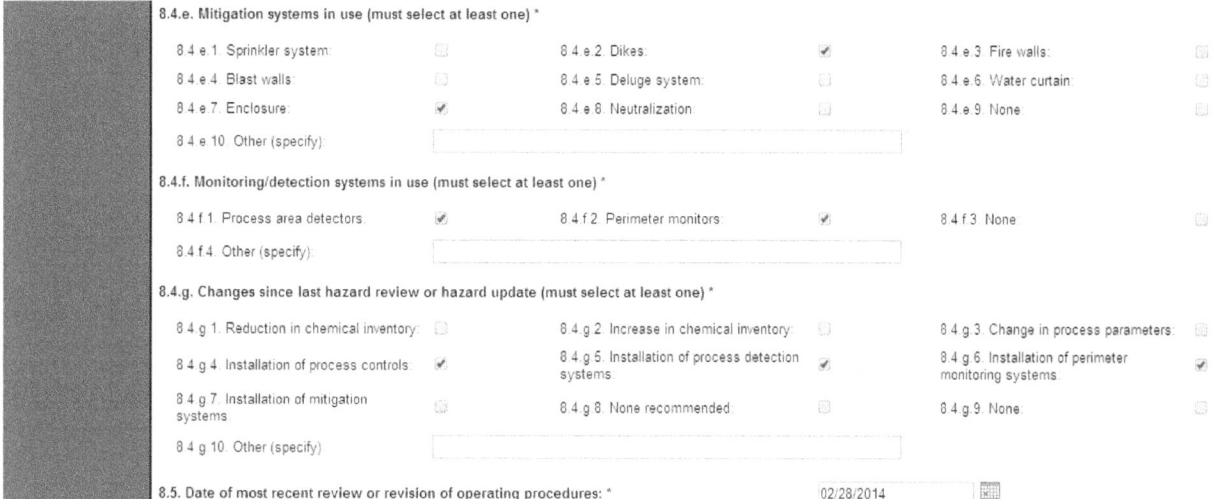

e. Mitigation systems in use:

Provide all mitigation systems you have in place to control a release should one occur from this Program 2 process or part of the process. Mitigation systems are described in 7.4 f. If none are applicable, check *None*.

f. Monitoring/detection systems in use:

Provide all monitoring and detection systems installed to detect a release of a regulated substance from the Program 2 process or part of the process. Monitoring and detection systems are described in 7.4 g. If none are applicable, check *None*.

g. Changes since last hazard review or hazard review update:

Provide all changes made to the process or part of the process since the last hazard review. Hazard review changes are described in 7.4 h. If none are applicable, check *None*.

8.5   Date of most recent review or revision of operating procedures:

The operating procedures requirements for Program 2 processes can be found in the RMP regulation at 40 CFR 68.52. For your first RMP, provide the date by which you complied with the requirements of Section 68.52(a) (prepare written procedures) for the process. For subsequent RMPs, provide the date of the most recent review or revision of the operating procedures. Operating procedures may be reviewed or revised as a result of, among other things, a major change to the process (see Section 68.52(c)) or a periodic audit of the prevention program (see Section 68.58(a)).(See Chapter 7 of the *General Guidance for Risk Management Programs* for a discussion of what constitutes a major change.) If the operating procedures were not reviewed or revised since the previous RMP was submitted, retain the date provided in the previous RMP.

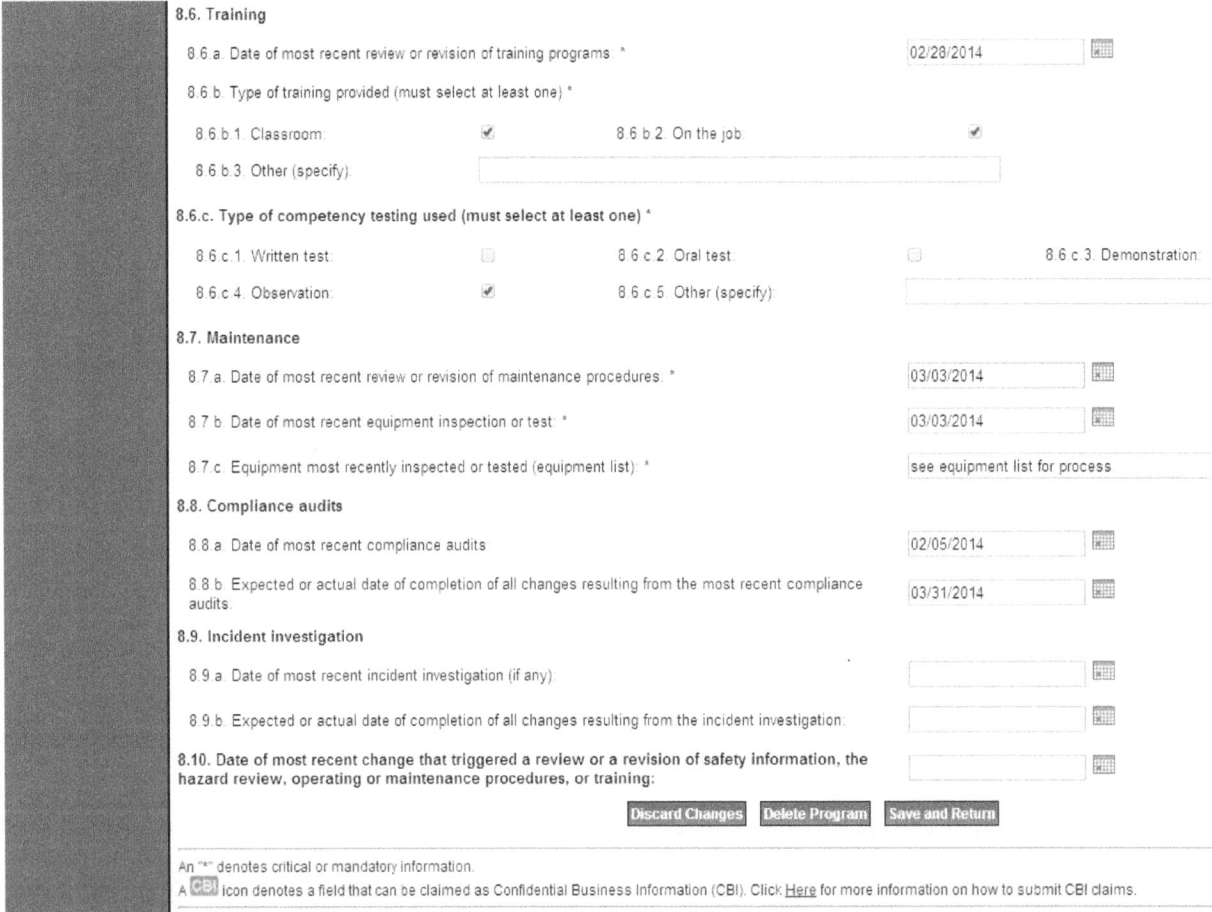

8.6   Training:

Training requirements for Program 2 processes can be found at 40 CFR 68.54.

a. Date of most recent review or revision of training programs:

For your first RMP, provide the date by which you ensured that the training you provide the employees operating the process meets the requirements of Section 68.54 (a). For subsequent RMPs, provide the date of the most recent review or revision of the training you provide. Training programs may be reviewed or revised as a result of, among other things, a change to the process (see 40 CFR 68.54 (d) (c)) or a periodic audit of the prevention program (see 40 CFR 68.58 (a)). If the training was not reviewed or revised since the previous RMP was submitted, retain the date provided in the previous RMP.

b. Type of training provided:

Select the type of training provided (select all that apply).

1. Classroom
2. On the job
3. Other (specify)

c. Type of competency testing used:

Indicate the type of competency test used. Competency tests are used to determine and evaluate comprehension of the training materials. Training information can be found in the RMP regulation at 40 CFR 68.54 and 68.71.

1. Written test
2. Oral test
3. Demonstration
4. Observation
5. Other (specify)

8.7 Maintenance:

a. The date that you most recently reviewed or revised the maintenance procedures:

For your first RMP, provide the date by which you complied with the requirements of Section 68.56 (a) (establish and implement written maintenance procedures) for the process. For subsequent RMPs, provide the date of the most recent review or revision of the maintenance procedures. Maintenance procedures may be reviewed or revised as a result of, among other things, a change to the process (see Section 68.56(a)) or a periodic audit of the prevention program (see Section 68.58(a)). If the procedures were not reviewed or revised since the previous RMP was submitted, retain the date provided in the previous RMP.

b. The date of the most recent equipment inspection or test:

Provide the appropriate date. Maintenance information can be found in the RMP regulation at 40 CFR 68.56 and 68.73.

c. The equipment most recently inspected or tested (list equipment):

Specify the equipment inspected or tested. Maintenance information can be found in the RMP regulation at 40 CFR 68.56 and 68.73.

8.8   Compliance audits:

Maintenance requirements for Program 2 processes can be found at 40 CFR 68.58. If you have not conducted a compliance audit prior to your first submission, leave these fields blank.

a. The date of your most recent compliance audit:

Provide the date of your most recent compliance audit.

NOTE: *A compliance audit is required every three years. Compliance audit information can be found in the RMP regulation at 40 CFR 68.58 and 68.79.*

b. The expected or actual date of completion of all changes resulting from the compliance audit:

Provide the expected or actual date of completion of all changes resulting from compliance audit. This may be left blank if there were no changes or all changes are complete. Incident Investigation information can be found in the RMP regulation at 40 CFR 68.60 and 68.81.

8.9   Incident investigation:

Maintenance requirements for Program 2 processes can be found at 40 CFR 68.60.

a. The date of your most recent incident investigation (if any):

Provide the date of your most recent incident investigation. Incident investigation (if any) can be found in the RMP regulation at 40 CFR 68.60. If you have not had an incident investigation, leave this field blank.

b. The expected or actual date of completion of all changes resulting from the incident investigation:

Provide the expected or actual date or completion of all changes resulting from the incident investigation. This may be left blank if there were no changes or all changes are complete. Incident Investigation information can be found in the RMP regulation at 40 CFR 68.60 and 68.81.

8.10   Date of most recent change that triggered review or revision:

Provide the date of most recent change that triggered review or revision of safety information, the operating or maintenance procedures, or training. This may be left blank if there were no changes.

Section 9: Emergency Response

The extent to which you need to fill out this portion of the RMP depends on whether your employees will respond to releases of regulated substances at your facility. Under Section 68.90(b), if your employees will *not* respond to releases, you are not required to comply with the requirements for an emergency response program provided you meet the following criteria:

1. If you hold one or more regulated toxic substances over threshold quantities, your facility must be included in the community emergency response plan developed under the Emergency Planning and Community Right-to-Know Act (EPCRA);

2. If you hold only one or more regulated flammable substances over threshold quantities, you must have coordinated response actions with the local fire department; and

3. You must have appropriate mechanisms in place to notify emergency responders when there is a need for a response.

If your employees will respond to releases of regulated substances at your facility, you are subject to Section 68.95 and must fill out all the data items in this section of the RMP. If your employees do not respond to releases of regulated substances at your facility, you need only respond to the first two (9.1 a and 9.1 b) and last three (9.7 a, 9.7 b and 9.8) emergency response data elements.

Complete this section once for all covered processes.

The following is a discussion of each element in *Section 9. Emergency Response.*

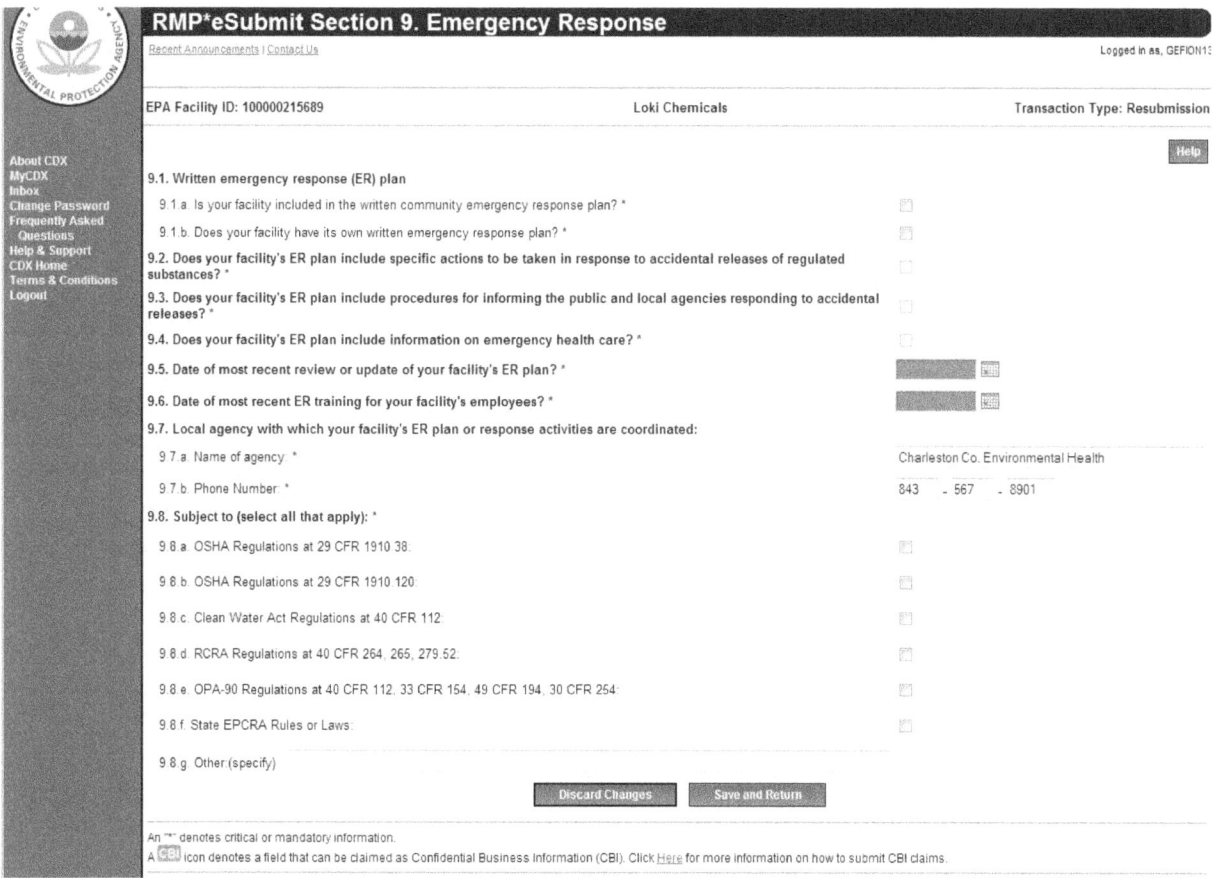

9.1   *Written emergency response (ER) plan:*

a. Is your facility included in the community emergency response plan?

If your facility is subject to part 68 because it has one or more regulated toxic substances above threshold quantities, it is probably included in a local emergency response plan under the Emergency Planning and Community Right-to-Know Act (EPCRA). Under Section 303 of EPCRA, local emergency planning committees (LEPCs) must prepare an emergency response plan for facilities in their planning district having toxic substances listed under EPCRA 302 in excess of the threshold planning quantity established under that section. Most of the toxic substances listed in Part 68 are also listed under EPCRA 302, and the EPCRA thresholds for those substances are generally the same or lower than the Part 68 thresholds for the same substances. Consequently, Part 68 facilities with toxic substances listed under both EPCRA and Part 68 should be included in community emergency response plans.

In addition, facilities subject to Part 68 as a result of flammable substances may also be covered by community emergency response plans, since LEPCs can, and sometimes do, include other hazardous substances, including flammables, in their plans. If you are not sure whether your facility is included in your community's local emergency plan, check with your LEPC.

As noted above, if your employees are *not* going to respond to releases of regulated substances at your facility and you have one or more Part 68 regulated toxic substances over threshold quantities, your facility must be included in the local emergency response place under EPCRA. Click the check box for this question if your facility is included in the community's emergency response plan.

b. Does your facility have its own written emergency response plan?

Click the check box for this question if you have a response plan (not just an emergency action plan as required by OSHA under 29 CFR 1910.38).

9.2　*Does your facility's ER plan include specific actions to be taken in response to accidental releases of regulated substances?*

These data elements (9.2, 9.3, 9.4) reflect the three mandatory components of the emergency response plan required under Section 68.95(a)(1). For an emergency response plan to be in compliance with this requirement, you must be able to answer "yes" to each of these questions. Click the check box for this question if your facility's ER plan includes specific actions to be taken in response to accidental releases of regulated substance(s).

9.3　*Does your facility's ER plan include procedures for informing the public and local agencies responding to accidental releases?*

Click the check box for this question if your facility's ER plan includes procedures for informing the public and local agencies responding to accidental releases.

9.4　*Does your facility's ER plan include information on emergency health care?*

Click the check box for this field to respond with "yes" to each of this question.

9.5   Date of most recent review or update of your facility's ER plan:

Provide the date on which you most recently reviewed or updated your plan. Section 68.95(a)(4) requires that ER plans be reviewed and updated "as appropriate" to reflect changes at the facility and to ensure that employees are informed of changes.

9.6   Date of the most recent ER training for your facility's employees:

Provide the date of the most recent emergency response training at your facility. Emergency response training includes drills involving your personnel with or without outside emergency response agencies and tabletop exercises of your emergency response plan. Single purpose drills (e.g., alarm system drills) may be listed, but exercises that test more aspects of the plan are preferable.

Part 68 does not specify a schedule for conducting employee response training. You should not, however, that other planning requirements (e.g., HAZWOPER) may establish a more formal schedule for conducting training (e.g., eight hours of annual refresher training.)

9.7   Local agency with which your facility's ER plan or response actions are coordinated:

If you have an ER plan, indicate the name and phone number of the agency with whom you have coordinated your plan. Section 68.95(c) requires that a facility's ER plan be coordinated with the community emergency response plan under EPCRA for the facility's community. The LEPC for the facility's community will typically be the agency with which ER plans are coordinated.

If you do not have an ER plan, indicate the agency with which you have coordinated response activities. As noted above, section 68.90(b) provides that if you have regulated toxic substances and your employees will not be responding to releases of those substances, your facility must be included in the community emergency response plan developed by the LEPC for your community. If that is the case for your facility, indicate the name and phone number of your LEPC here. If you have only regulated flammable substances and your employees will *not* be responding to releases of those substances, you must have coordinated response actions with the local fire department. If that is case for your facility, indicate the name and phone number of your local fire department here.

    a. Name of agency:

    If you have an ER plan, provide the name of the agency with whom you have coordinated your plan.

    b. Phone number:

    If you have an ER plan, provide the phone number of the agency with whom you have coordinated your plan.

9.8   Subject to (select all that apply):

Indicate all of the federal and state emergency response regulations or statutes to which your facility is subject. Select at least one. All RMP facilities are subject to OSHA emergency planning requirements at 29 CFR 1910.38 or 29 CFR 1910.120.

   a. **OSHA Regulations at 29 CFR 1910.38.** These are OSHA's Emergency Action Plan regulations. All RMP facilities are subject to either these OSHA regulations or OSHA regulations at 29 CFR 1910.120.

   b. **OSHA Regulations at 29 CFR 1910.120.** These are OSHA's Hazardous Waste Operations and Emergency Response (HAZWOPER) Plan regulations. All RMP facilities are subject to either these OSHA regulations or OSHA regulations at 29 CFR 1910.38.

   c. **Clean Water Act Regulations at 40 CFR 112.** These are EPA's Oil Spill Prevention Control and Countermeasures (SPCC) regulations under the Clean Water Act.

   d. **RCRA Regulations at 40 CFR 264, 265, and 279.52.** These are EPA's permitting regulations for solid waste under the Resource Conservation and Recovery Act (RCRA).

   e. **OPA 90 Regulations at 40 CFR 112, 33 CFR 154, 49 CFR 194, or 30 CFR 254.** These are EPA, U.S. Coast Guard, Department of Transportation, and Department of the Interior facility response plan regulations under the Oil Pollution Act of 1990 (OPA 90).

   f. **State EPCRA Rules or Laws.** These are state emergency planning and community right-to-know (EPCRA) laws. Federal EPCRA does not require facility response plans, but some state laws may.

   g. **Other.** Specify any other emergency response regulations or laws to which your facility is subject.

Executive Summary

The *Executive Summary* must include a brief description of your facility's risk management program. You determine the length; it may be as short as two or three pages or, if you have many processes, it may need to be longer. You should view the *Executive Summary* as an opportunity to communicate in your own words the nature of the risks posed by your facility to your community and to explain what you have done to minimize those risks. The summary can be an excellent vehicle to display the effort and resources your facility has put into its accident prevention program. Your *Executive Summary* cannot be claimed as CBI. ***Do not include any <u>CBI</u> data in your Executive Summary.***

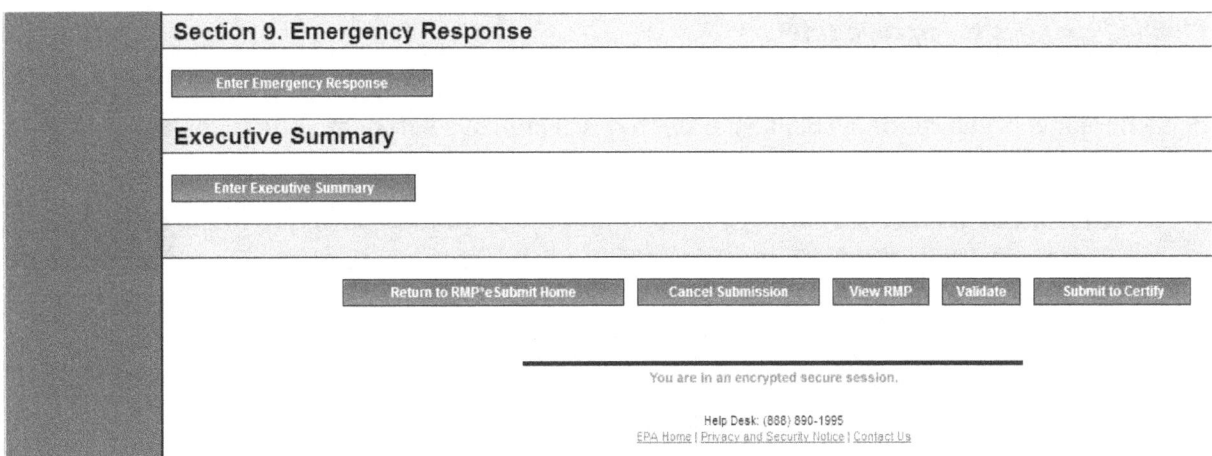

The following is a discussion of each element in *Executive Summary*.

The *Executive Summary* must briefly describe the following elements:

The accidental release prevention and emergency response policies at your facility

Describe your facility's overall approach to chemical safety. You may want to include any corporate policies (if applicable) and an overview of senior management commitment to safety and implementation of safe procedures.

Your facility and the regulated substances handled

Provide a description of your facility so that the public has a clear picture of the facility, its processes, and products. Describe the primary activities at the facility (e.g., manufacturer of polyethylene, pulp mill, etc.) and the regulated substances used. In addition, you may want to mention the quantities of these substances handled or stored at your facility.

The general accidental release prevention program and chemical-specific prevention steps

You may wish to mention the rules and regulations with which your facility complies, such as the OSHA PSM rule. You should also highlight practices that you believe are important to your prevention program. The steps you list may be either technological (e.g., backup systems) or procedural/managerial (e.g., improved maintenance or training).

The five-year accident history

Do not present accident history information in table form here; more details will be provided in the data elements. This should be a written summary; for example,

> *"We have had five accidental releases of chlorine in the past five years; the largest release was 1,500 pounds. No one offsite was injured, but several houses were evacuated as a precautionary measure during the October 2005 and May 2006 releases."*

The emergency response program

Briefly describe the elements of your response program. These may include coordination with local emergency responders, training received by personnel, drills conducted by your facility, public notification and alert systems, as appropriate.

Planned changes to improve safety

List any upcoming events, such as training, installation of new mitigation or control equipment or technology, organizational changes, etc., that will improve safety at your facility.

> ***NOTE: A summary of the off-site consequence analysis (OCA) for the worst-case and alternative release scenarios(s) is no longer required to be included in the Executive Summary.*** *While the RMP rule originally required that the Executive Summary briefly describe the OCA for worst-case and alternative release scenario(s), EPA amended the RMP rule in 2004 to remove this requirement because of security concerns. Your Executive Summary should not describe nor include information concerning your worst case or alternative release scenarios.*

After you have finished entering your *Executive Summary*, click the *Save and Return* button. This will take you back to the RMP*eSubmit Section Selection page.

View RMP, Validate, Submit to Certify

At the bottom of the RMP*eSubmit Section Selection page, click the *Validate* button to view and adjust errors to your RMP before submitting it for the Certifying Official for review. Once any errors have been addressed, you can click the *Submit to Certify* button to send the RMP to the Certifying Official. Your next page should be similar to the screen shot below.

CHAPTER 3 HOW TO SUBMIT YOUR RMP

Certifying a Submission in RMP*eSubmit

After the Preparer has prepared and submitted the RMP to the Certifying Official, the Certifying Official (Certifier) approves the RMP by certifying and submitting to EPA.

> **WARNING! Your RMP has not been officially submitted to EPA until it has been certified and submitted by the Certifying Official. Please be sure to take the step of certifying your RMP. A confirmation email will be sent to acknowledge the submission.**
>
> **<u>Note to Preparer/Certifier</u>: If you have the roles of both Preparer and Certifier, please be sure to take the step beyond submitting to Certifier. You, as the Certifying Official, must certify the submission.**

The Certifier may reject and send the RMP back to the Preparer for possible corrections.

An automated email will be sent to the Certifier and the Preparer. This email is a notification that a submission has been made by the Preparer and is ready for review and submission to EPA by the Certifier. The Certifier's *MyCDX* "Certify Submission" link will show a list of Pending Submissions to be certified.

The Certifier can login to *MyCDX* and either (a) **Certify** or (b) **Reject** a submitted RMP. The Certifier may also review a previously submitted, certified, and approved RMP for which the Certifier has authorized access.

If the submission is rejected, an email will be sent to the Preparer to make necessary corrections before they submit the RMP again to the Certifier for certification and submission.

The Certifier has the option to select and review a copy of the previously submitted approved RMP for a facility by clicking the "View Current RMP" on the list of associated RMPs.

About CDX
MyCDX
Inbox
Change Password
Frequently Asked
   Questions
Help & Support
CDX Home
Terms & Conditions
Logout

Here is a list of RMP submissions that have been prepared for you to certify and send to EPA. Please review the information in these submissions, by clicking a File ID link, and certify the submissions.

Submitter	Facility ID	CDX ID/File ID	
Jacob Noble (JPNRMPESUBMIT)	Calhoun County Water Filtration Plant (1000 0003 2975)	RMP000120090306174840JPNRMPESUBMIT	Certify	Reject

Non-Pending Submissions

Here is a list of RMP submissions that have been acted on by you. Click a File ID link to see the PDF of the submission. Click a Status link to see the receipt page of a submission.

Submitter	Facility ID	CDX ID/File ID	Status
Jacob Noble (JPNRMPESUBMIT)	Calhoun County Water Filtration Plant (1000 0003 2975)	RMP000120090306173146JPNRMPESUBMIT	Certified

Facilities

Your CDX account is associated with one or more Facilities IDs:

Facility ID	Facility Name	View Currently Approved RMP
1000 0003 2975	Calhoun County Water Filtration Plant	View Current RMP
1000 0013 9094	Woodland Biomass Power Ltd.	View Current RMP

The RMP selected to be viewed will be in PDF format. See the example below.

Section 1. Registration Information

1.1 Source Identification	
1.1.a. Facility Name	Calhoun County Water Filtration Plant
1.1.b. Parent Company #1 Name	Calhoun County Water Authority
1.1.c. Parent Company #2 Name	
1.2 EPA Facility Identifier	100000032975
1.3 Other EPA Systems Facility Identifier	
1.4 Dun and Bradstreet Numbers (DUNS)	
1.4.a. Facility DUNS	
1.4.b. Parent Company #1 DUNS	
1.4.c. Parent Company #2 DUNS	
1.5 Facility Location	
1.5.a. Street - Line 1	630 Smith Boozer Road
1.5.b. Street - Line 2	
1.5.c. City	Wellington
1.5.d. State	AL
1.5.e. Zip Code - Zip +4 Code	36279-5829
1.5.f. County	CALHOUN
1.5.g. Facility Latitude (in decimal degrees)	33.864630
1.5.h. Facility Longitude (in decimal degrees)	-085.883850
1.5.i. Method for determining Lat/Long	Interpolation - Map
1.5.j. Description of location identified by Lat/Long	Center of Facility
1.5.k. Horizontal Accuracy Measure (meters)	25
1.5.l. Horizontal Reference Datum Code	World Geodetic System of 1984
1.5.m. Source Map Scale Number	1
1.6 Owner or Operator	

If the Certifier accepts and proceeds with the submission, the Certifier will follow the next process to electronically sign the certification statement.

Certification Statement (Part 1)

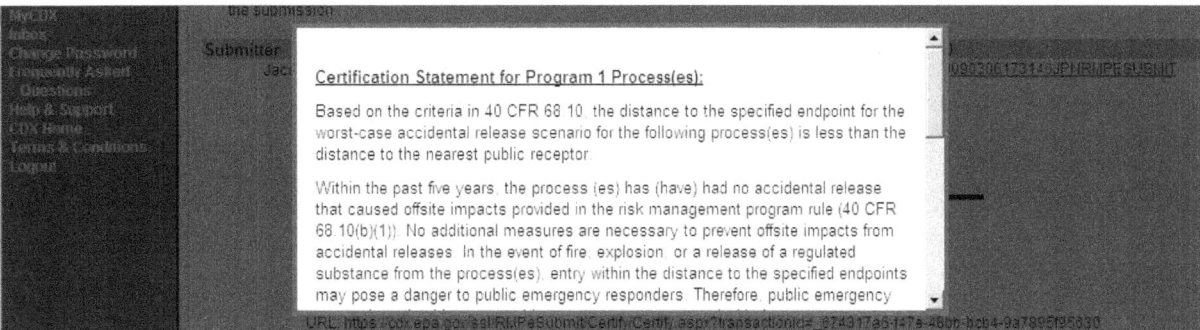

Certification Statement (Part 2)

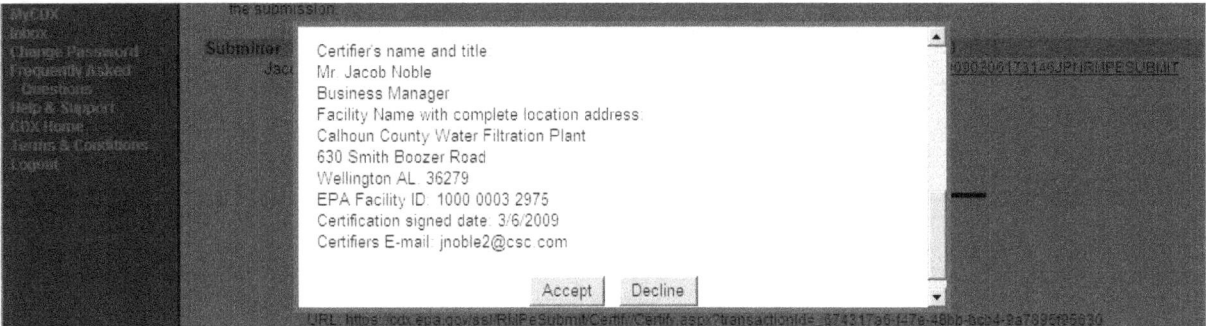

Click the "Accept" button to complete certification, or click the "Decline" button.

Authentication of CDX Credentials

Enter your CDX password to authenticate the submission and click "Login".

Re-authentication of CDX Credentials

Now the user will be challenged with 1 of the 5 questions and answers in the registration process. This step will re-authenticate your CDX credentials. Answer the question, then click the "Answer" button.

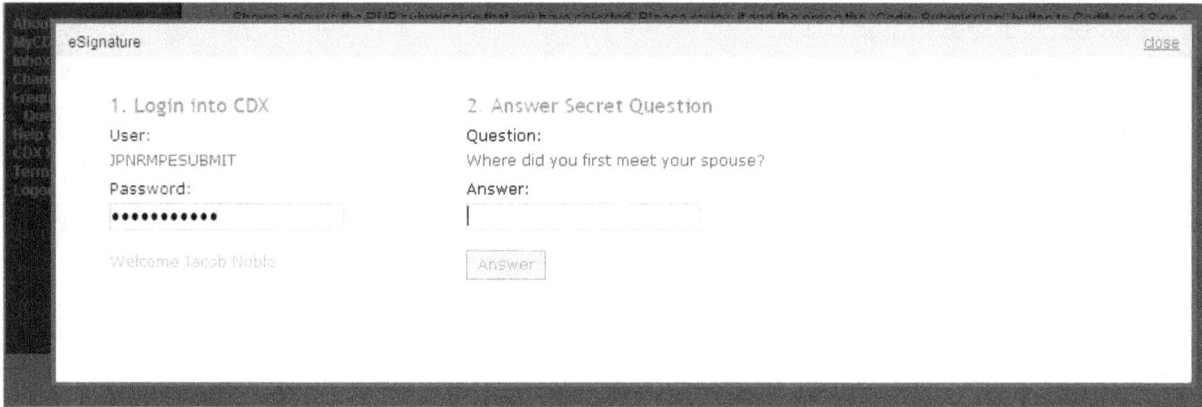

Electronic Signature

Click the "Sign" button to submit your RMP.

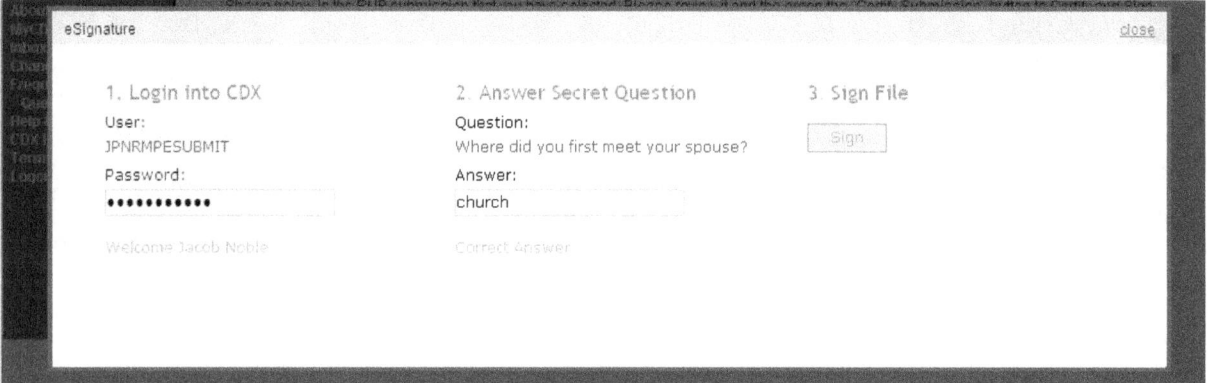

The Certifier will see the following screen as an acknowledgement for the submission.

CHAPTER 4 SUBMITTING CONFIDENTIAL BUSINESS INFORMATION AND PAPER RMPS

Confidential Business Information

On January 6, 1999, EPA published a final rule in the Federal Register specifying which RMP data elements may not be claimed as confidential business information (CBI) and the procedures which must be followed to claim information as CBI. The regulatory provisions (Sections 68.151 and 68.152), provide that if you claim any RMP information as CBI, you must submit to EPA a sanitized RMP, a Substantiation Form (for explaining why you believe the information meets the criteria for CBI), and an Unsanitized Data Elements Form (on paper only) (see Appendices C and D for forms). **The sanitized (also known as redacted) RMP <u>should not include any confidential business information.</u>** If you must claim any CBI in your RMP, click the button after the question "Are you claiming Confidential Business Information (CBI) in this Section?" at the top of each section. Please read the pop-up Warning Message.

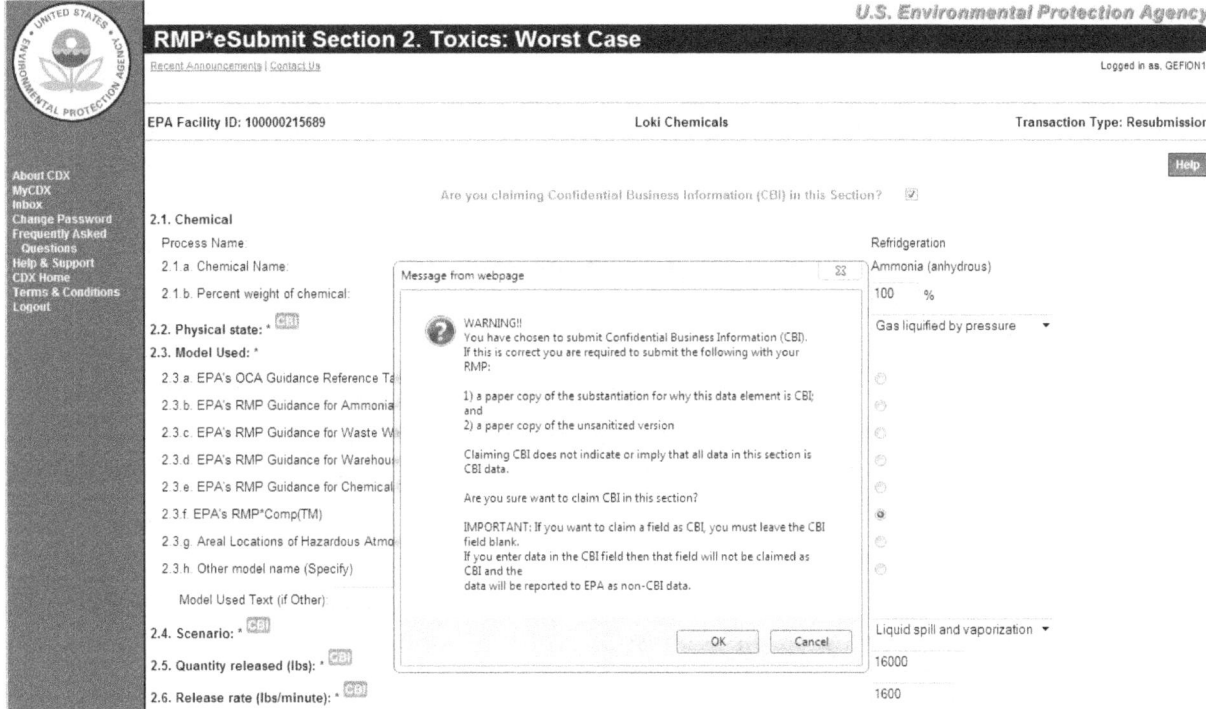

For additional information, contact the RMP Reporting Center.

Submitting Paper RMPs

If you are unable to submit your RMP online, use Appendix A: Risk Management Plan Form in this manual or call the RMP Reporting Center for assistance.

CHAPTER 5 NOTIFYING EPA THAT YOUR FACILITY IS NO LONGER COVERED BY RMP: DE-REGISTRATION

How to De-register Your Facility

Changes may occur at your facility that make it no longer subject to the RMP regulations at 40 CFR Part 68 (e.g., you replace the regulated substances in your process with unregulated substances.) If your facility is no longer covered by RMP, you must notify EPA as specified in Section 68.190(c) (see box below). Note that the regulation uses the term "stationary source" to refer to a facility.

68.190 Updates

(c) If a stationary source is no longer subject to this part, the owner or operator shall submit a de-registration to EPA within six months indicating that the stationary source is no longer covered.

To de-register, submit a letter to the RMP Reporting Center within six months and include the effective date of the de-registration (the date on which your facility was no longer covered by part 68). The letter is to be signed by the owner or operator and include your RMP ID number (the 12-digit ID number assigned by EPA).

Use the Risk Management Program De-registration Form (Appendix E) of this manual.

If you de-register your facility and it later again becomes subject to the RMP regulation, you will resubmit an RMP following the resubmission process. You must use the original EPA Facility ID # to resubmit. Keep a record of your Facility ID # upon de-registering. If you have de-registered and do not have your Facility ID #, contact the RMP Reporting Center.

> ***Important Reminder:*** *Remember to include the 12-digit EPA Facility Identification number (usually beginning with 1000) that was originally assigned to your facility. The EPA Facility ID was given to you in the notification letter you received from the RMP Reporting Center regarding the submission status of your RMP.*

CHAPTER 6 PROPANE WITHDRAWAL FORM AND REQUEST TO CONSOLIDATE EPA FACILITY IDS FORM

When to Uses these Forms

If you have submitted an RMP and your facility is no longer covered under 40 CFR part 68 because the facility does not have more than a threshold quantity of any regulated substance listed at 40 CFR 68.130 other than propane (or another listed flammable substance) that is used on site as fuel or held for sale as fuel at a retail facility (see 40 CFR 68.126), use the Propane Facility Withdrawal Form, Appendix F.

If your facility has inadvertently been assigned more than one EPA Facility ID number, use the Request to Consolidate EPA Facility IDs Form, Appendix G, to request the EPA Facility ID numbers be consolidated into a single ID number. This form can also be used to de-activate additional/incorrect EPA Facility ID numbers.

CHAPTER 7 REPORTING CENTER CONTACT INFORMATION ACRONYMS OF TERMS

RMP Reporting Center Contact Information

U.S. Environmental Protection Agency
Attention: RMP Reporting Center
P.O. Box 10162
Fairfax, VA 22038

For courier and overnight delivery packages, use the address below:

RMP Reporting Center
c/o CGI Federal, Inc.
12601 Fair Lakes Circle
Fairfax, VA 22033

Acronyms of Terms

ANSI	American National Standards Institute Standards
ASTM	American Society of Testing Materials Standards
AUTHCODE	Authorization Code
BLEVE	Boiling Liquid Expanding Vapor
CAA	Clean Air Act
CAS	Chemical Abstracts Service Registry Number
CBI	Confidential Business Information
CDX	Central Data Exchange
CERCLA	Comprehensive Environmental Response, Compensation, and Liability Act
DUNS	Data Universal Numbering System
EHS	Extremely Hazardous Substance
EPA	Environmental Protection Agency
EPCRA	Emergency Planning and Community Right-to-Know Act
ESA	Electronic Signature Agreement
FII	Facility Identification Number
GIS	Geographic Information System
GPS	Global Positioning System
HAZWOPER	Hazardous Waste Operations and Emergency Response
LEPC	Local Emergency Planning Committee
MSDS	Material Safety Data Sheet
NAD	North American Datum
NAICS	North American Industrial Classification System
NCDC	National Climatic Data Center
NFPA	National Fire Protection Association
OCA	Offsite Consequence Analysis
OPA	Oil Pollution Act
OSHA	Occupational Safety and Health Administration
PHA	Process Hazard Analysis
PIN	Personal Identification Number
POTWs	Publicly Owner Treatment Works
PSM	Process Safety Management
RCRA	Resource Conservation and Recovery Act
RMP	Risk Management Plan
TRI	Toxic Release Act
UIN	Unique Identification Number
WGS	World Geodetic System

APPENDICES

APPENDIX A. RISK MANAGEMENT PLAN FORM

Risk Management Plan Form
Section 112(r) of the Clean Air Act

Control Number 2050-0144

IMPORTANT: *Type or print; read instructions before completing form.*

Submission Type:	**Where to Send Completed Forms:**
❏ First-Time RMP Submission	**U.S. Environmental Protection Agency**
	Attention:RMP Reporting Center
❏ Correction to the Current RMP	**P.O. Box 10162**
	Fairfax, VA 22038
(Submission Type = "C")	
	If you prefer to send this Risk Management Plan
	Form by certified mail, courier or overnight mail

(Submission Type = "C")

C01	Clerical error corrected	
C02	Additional information supplied	
C03	Minor administrative change	
C04	Notification of facility ownership change	
C05	New accident history information	
C06	Change in emergency contact information	
C07	New data element required by EPA	
C08	Optional data element requested by EPA	
C09	Removed OCA description from executive summary	

If you prefer to send this Risk Management Plan Form by certified mail, courier or overnight mail (e.g. Fed Ex, UPS, Etc.), please address it to:

RMP Reporting Center
c/o CGI Federal, Inc.
12601 Fairlakes Circle
Fairfax, VA 22033

❏ Re-Submission (all 9 sections are updated and certified)

(Submission Type = "R")

R01	Newly regulated substance listed by EPA (40 CFR 68.190(b)(2))
R02	Newly regulated substance above TQ in already covered process (40 CFR 68.190(b)(3))
R03	Regulated substance present above TQ in new (or previously not covered) process (40 CFR 68.190(b)(4))
R04	Revised PHA / Hazard Review due to process change (40 CFR 68.190(b)(5))
R05	Revised OCA due to change (40 CFR 68.190(b)(6))
R06	Change in program level of covered process (40 CFR 68.190(b)(7))
R07	5-year update (40 CFR 68.190(b)(1))
R08	Process no longer covered (source has other processes that remain covered) (40 CFR 68.190(b)(7))
R09	Voluntary update (not described by any of the above reasons)

EPA Form 8700-25   -1-

Facility Name: _____

ES

Executive Summary
(attach a separate piece of paper if you need additional space)

⌐_,_,_,_,⌐ - ⌐_,_,_,_,⌐ - ⌐_,_,_,_⌐
EPA Facility ID# (leave blank for first submission only)

EPA Form 8700-25   -2-

Facility Name: _____

1 **Section 1. Registration**	⌐__,__,__,__,__¬ - ⌐__,__,__,__,__¬ - ⌐__,__,__,__¬ EPA Facility ID# (leave blank for first submission only)

1.1 Source Identification

1.1.a. Facility Name (maximum 50 characters)
1.1.b. Parent Company #1 Name (maximum 50 characters)
1.1.c. Parent Company #2 Name (maximum 50 characters)

1.2 EPA Facility Identifier (12 characters)   ⌐__,__,__,__,__¬ - ⌐__,__,__,__,__¬ - ⌐__,__,__,__¬   (leave blank for first submission only)

1.3 Other EPA Systems Facility Identifier (15 characters)   ⌐__,__,__,__,__,__,__,__,__,__,__,__,__,__,__¬

1.4 Dun and Broadcast Numbers (DUNS) (9 characters)

1.4.a. Facility DUNS	1.4.b Parent Company #1 DUNS	1.4.c. Parent Company #2 DUNS
⌐__,__,__,__,__,__,__,__,__¬	⌐__,__,__,__,__,__,__,__,__¬	⌐__,__,__,__,__,__,__,__,__¬

1.5 Facility Location

1.5.a. Street - Line 1 (maximum 35 characters)	
1.5.b. Street - Line 2 (maximum 35 characters)	
1.5.c. City (maximum 19 Characters)	1.5.d. State ⌐__,__¬
1.5.e. Zip Code Zip +4 Code ⌐__,__,__,__,__¬ ⌐__,__,__,__¬	1.5.f. County (maximum 20 characters)
1.5.g. Facility Latitude (report in decimal degrees) ⌐_¬ ⌐__,__,__¬ . ⌐__,__,__,__¬ ⌐_¬ +/- D DD D DDDD D	1.5.h. Facility Longitude (report in decimal degrees) ⌐_¬ ⌐__,__,__¬ . ⌐__,__,__,__¬ ⌐_¬ +/- D DD D DDDD D
1.5.i. Method for determining Lat/Long (see User Manual for Codes) ⌐__,__¬	1.5.j. Description of location identified by Lat/Long (see User Manual for Codes) ⌐__,__¬
1.5.k. Horizontal accuracy measure (meters) ⌐__,__,__¬	1.5.l. Horizontal reference datum code ⌐__,__,__¬
	1.5.m. Source Map Scale Number _____

Facility Name: _____

1　**Section 1. Registration**	⎕⎕⎕⎕ - ⎕⎕⎕⎕⎕ - ⎕⎕⎕⎕⎕ **EPA Facility ID#** (leave blank for first submission only)

1.6 Owner or Operator

1.6.a. Name (maximum 35 characters)

1.6.b. Phone　(⎕⎕⎕) ⎕⎕⎕ - ⎕⎕⎕⎕

Owner or Operator Mailing Address

1.6.c. Street - Line 1 (maximum 35 characters)

1.6.d. Street - Line 2 (maximum 35 characters)

1.6.e. City (maximum 19 characters)	1.6.f. State ⎕⎕
1.6.g. Zip Code　　Zip +4 Code ⎕⎕⎕⎕⎕　　⎕⎕⎕⎕	

1.7 Name, title, and email address of person or position responsible for RMP (part 68) implementation

1.7.a. Name of person (maximum 35 characters)	1.7.b. Title of person or position (maximum 35 characters)
1.7.c. Email address of person or position (maximum 35 characters)	

1.8.a. Emergency Contact

1.8.a. Name (maximum 35 characters)	1.8.b. Title of person or position (maximum 35 characters)
1.8.c. Phone (⎕⎕⎕) ⎕⎕⎕ - ⎕⎕⎕⎕	1.8.d. 24-Hour Phone (⎕⎕⎕) ⎕⎕⎕ - ⎕⎕⎕⎕
1.8.e. 24-Hour Phone Extension/PIN # (maximum 10 characters)	
1.8.f. Email address for emergency contact (maximum 100 characters)　　*Enter N/A if not applicable*	

EPA Form 8700-25　　　　　　-4-

Facility Name: _____

EPA Facility ID# (leave blank for first submission only)

1   **Section 1. Registration**

1.9. Other Points of Contact (Optional)

1.9.a. Facility or Parent Company E-mail Address (maximum 100 characters)	1.9.b. Facility Public Contact Phone Number
1.9.c. Facility or Parent Company WWW Homepage Address (maximum 100 characters)	

1.10 Local Emergency Planning Committee (LEPC) (optional) (maximum 30 characters)

1.11 Number of full-time equivalent (FTEs) employees on site

1.12. Covered by (select all that apply)

❑ 1.12.a. OSHA PSM

❑ 1.12.b. EPCRA section 302

❑ 1.12.c. CAA Title V Air Operating Permit Program. If covered, specify permit ID# below.

1.13. OSHA Star or Merit Ranking (optional)   ❑ YES   ❑ NO

1.14. Last Safety Inspection (by an External Agency) Date   M M   D D   Y Y Y Y

1.15. Last Safety Inspection Performed by an External Agency (select one)

❑ 1.15.a. OSHA
❑ 1.15.b. State occupational safety agency
❑ 1.15.c. EPA
❑ 1.15.d. State Environmental Agency
❑ 1.15.e. Fire Department

❑ 1.15.f. Never had one
❑ 1.15.g. Other (specify) (maximum 50 characters)

_____

_____

1.16. Will this RMP involve Predictive Filing? (Optional)   ❑ YES   ❑ No

EPA Form 8700-25   -5-

Facility Name: _____

┗━━┛ - ┗━━┛ - ┗━━┛
EPA Facility ID# (leave blank for first submission only)

1 Section 1. Registration

1.17 Process Specific Information. For each covered process, fill in this page. If you are reporting more than one process, make a photocopy of this page and report each process on a separate sheet.

Process ID# (optional - for your reference only)

Process Description (optional - for your reference only)

1.17.a. Program Level (select one)   ❑ 1   ❑ 2   ❑ 3

1.17.b. NAICS Code(s) (five or six digits)

┗━━━━━┛   ┗━━━━━┛   ┗━━━━━┛   ┗━━━━━┛

1.17.c. Chemical(s) (regulated substance(s))

1.17.c.1. Name (maximum 100 characters)	1.17.c.2. CAS Number (10 characters)	1.17.c.3. Quantity (lbs) (max. 12 chars.)
	┗━━━━━┛	
	┗━━━━━┛	
	┗━━━━━┛	
	┗━━━━━┛	
	┗━━━━━┛	
	┗━━━━━┛	
	┗━━━━━┛	
	┗━━━━━┛	
	┗━━━━━┛	

If you need more space to list NAICS codes or chemicals, please make a photocopy of this sheet.

EPA Form 8700-25   -6-

Facility Name: _____

⌐_ _ _ _¬ − ⌐_ _ _ _¬ − ⌐_ _ _ _¬

EPA Facility ID# (leave blank for first submission only)

1	**Section 1. Registration** If an outside contractor prepared this risk management plan, please enter information concerning this contractor in the fields below.

1.18 RMP Preparer Information

1.18.a. Name (maximum 70 characters)

1.18.b. Phone (⌐_ _ _ _¬) ⌐_ _ _ _¬ − ⌐_ _ _ _ _¬

1.18.c. Street - Line 1 (maximum 35 characters)

1.18.d. Street - Line 2 (maximum 35 characters)

1.18.e. City (Maximum 30 characters)

1.18.f. State ⌐_ _¬ or Foreign State or Province (Maximum 35 characters)	1.18.g. Zip Code Zip+ 4 Code ⌐_ _ _ _ _¬ − ⌐_ _ _ _¬	or Foreign Country (Max 2 characters)
1.18.h. RMP Preparer Foreign Zip Code		

EPA Form 8700-25   -7-

Facility Name: _____

EPA Facility ID# (leave blank for first submission only)

⎵⎵⎵⎵ - ⎵⎵⎵⎵ - ⎵⎵⎵⎵

2   **Section 2. Toxics: Worst Case**

(If you need to report more than one worst case scenario, make a photocopy of pages in this section and report each scenario separately)

2.1. Chemical

2.1.a. Name (maximum 100 characters)

2.1.b. Percent weight of chemicals (if in a mixture)   ⎵⎵ . ⎵ %

2.2. Physical state (select one)

❑ 2.2.a. Gas
❑ 2.2.b. Liquid

❑ 2.2.c. Gas liquified by pressure
❑ 2.2.d. Gas liquified by refrigeration

2.3. Model Used (select one or enter another model name in Other below)

❑ 2.3.a. EPA's OCA Guidance Reference Tables or Equations
❑ 2.3.b. EPA's RMP Guidance for Ammonia Refrigeration Reference Tables or Equations
❑ 2.3.d. EPA's RMP Guidance for Waste Water Treatment Plants Reference Tables or Equations
❑ 2.3.e. EPA's RMP Guidance for Warehouses Reference Tables or Equations
❑ 2.3.f. EPA's RMP Guidance for Chemical Distributors Reference Tables or Equations
❑ 2.3.g. EPA's RMP* Comp™
❑ 2.3.h. Areal Locations of Hazardous Atmospheres (ALOHA®)
❑ 2.3.z. Other model (specify) (maximum 255 characters)

_____

2.4. Scenario (select one)

❑ 2.4.a. Gas Release   ❑ 2.4.b. Liquid Spill and Vaporization

2.5. Quantity released (lbs)

⎵⎵⎵⎵⎵⎵⎵⎵⎵⎵

2.6. Release rate (lbs/minute)

⎵⎵⎵⎵⎵⎵⎵ . ⎵

2.7. Release duration (minutes)

⎵⎵⎵⎵ . ⎵

2.8. Wind speed (meters/second)

⎵⎵⎵⎵ . ⎵

2.9. Atmospheric stability class (A-F)

⎵

2.10. Topography (select one)

❑ 2.10.a. Urban   ❑ 2.10.b. Rural

2.11. Distance to endpoint (miles)

⎵⎵⎵ . ⎵⎵

EPA Form 8700-25   -8-

Facility Name: ______________________

2 **Section 2. Toxics: Worst Case**

___ - ___ - ___
EPA Facility ID# (leave blank for first submission only)

2.12. Estimated residential population within distance to endpoint (numeric)

___ , ___ , ___

2.13. Public receptors within distance to endpoint (select all that apply)

- ❑ 2.13.a. Schools
- ❑ 2.13.b. Residences
- ❑ 2.13.c. Hospitals
- ❑ 2.13.d. Prison/Correctional Facilities
- ❑ 2.13.e. Recreation Areas
- ❑ 2.13.f. Major commercial, office, or industrial areas
- ❑ 2.13.g. Other (specify) (maximum 200 characters)

2.14. Environmental receptors within distance to endpoint (select all that apply)

- ❑ 2.14.a. National or State Parks, Forests, or Monuments
- ❑ 2.14.b. Officially Designated Wildlife Sanctuaries, Preserves, or Refuges
- ❑ 2.14.c. Federal Wilderness Area
- ❑ 2.14.d. Other (specify) (maximum 200 characters)

2.15. Passive mitigation considered (select all that apply)

- ❑ 2.15.a. Dikes
- ❑ 2.15.b. Enclosures
- ❑ 2.15.c. Berms
- ❑ 2.15.d. Drains
- ❑ 2.15.e. Sumps
- ❑ 2.15.f. Other (specify) (maximum 200 characters)

2.16. Graphics file name (optional) (maximum 12 characters)

EPA Form 8700-25 -9-

Facility Name: _____

3	**Section 3. Toxics: Alternative Release**

EPA Facility ID# (leave blank for first submission only)

(If you need to report more than one alternative release scenario, make a copy of pages in this section and report each scenario separately)

3.1. Chemical

3.1.a. Name (maximum 100 characters)

3.1.b. Percent weight of chemical (if in a mixture)

___ . ___%

3.2. Physical State (select one)

❑ 3.2.a. Gas
❑ 3.2.b. Liquid
❑ 3.2.c. Gas liquified by pressure
❑ 3.2.d. Gas liquified by refrigeration

3.3. Model Used (select one or enter another model name in Other below)

❑ 3.3.a. EPA's OCA Guidance Reference Tables or Equations
❑ 3.3.b. EPA's RMP Guidance for Ammonia Refrigeration Reference Tables or Equations
❑ 3.3.d. EPA's RMP Guidance for Waste Water Treatment Plants Reference Tables or Equations
❑ 3.3.e. EPA's RMP Guidance for Warehouse Reference Tables or Equations
❑ 3.3.f. EPA's RMP Guidance for Chemical Distributors Reference Tables or Equations
❑ 3.3.g. EPA's RMP*Comp™
❑ 3.3.h. Areal Locations of Hazardous Atmospheres (ALOHA®)
❑ 3.3.z. Other model (specify) (maximum 200 characters)

_____

3.4. Scenario (select one)

❑ 3.4.a. Transfer hose failure
❑ 3.4.b. Pipe Leak
❑ 3.5.c. Vessel Leak
❑ 3.4.d. Overfilling
❑ 3.4.e. Rupture disk/relief valve failure

❑ 3.4.f. Excess Flow Device Failure
❑ 3.4.g. Other (specify) (maximum 35 characters)

_____

_____

3.5. Released (lbs)	3.6. Release Rate (lbs/minute)
3.7. Release Duration (minutes) ___ . ___	3.8. Wind Speed (meters/second) ___ . ___
3.9. Atmospheric stability class (A-F) ___	

EPA Form 8700-25   -10-

Facility Name: _____

3	**Section 3. Toxics: Alternative Release**

EPA Facility ID# (leave blank for first submission only)

(If you need to report more than one alternative release scenario, make a copy of pages in this section and report each scenario separately)

3.10. Topology (select one)

❏ 3.10.a. Urban   ❏ 3.10.b. Rural

3.11. Distance to endpoint (miles)

3.12. Estimated residential population within distance to endpoint

---

3.13. Public receptors within distance to endpoint (select all that apply)

❏ 3.13.a. Schools
❏ 3.13.b. Residences
❏ 3.13.c Hospitals
❏ 3.13.d. Prisons/Correctional facilities

❏ 3.13.e. Recreation Areas
❏ 3.13.f. Major commercial, office, or industrial areas
❏ 3.13.g. Other (specify) (maximum 200 characters)

3.14. Environmental receptors within distance to endpoint (select all that apply)

❏ 3.14.a. National or State Parks, Forests, or Monuments
❏ 3.14.b. Officially Designated Wildlife Sanctuaries, Preserves, or Refuges
❏ 3.14.c. Federal Wilderness Area

❏ 3.14.d. Other (specify) (maximum 200 characters)

3.15. Passive mitigation considered (select all that apply)

❏ 3.15.a. Dikes

❏ 3.15.b. Enclosures

❏ 3.15.c. Berms

❏ 3.15.d. Drains

❏ 3.15.e. Sumps

❏ 3.15.f. Other (specify) (maximum 200 characters)

3.16. Active mitigation considered (select all that apply)

❏ 3.16.a. Sprinkler systems
❏ 3.16.b. Deluge systems
❏ 3.16.c. Water curtain
❏ 3.16.d. Neutralization
❏ 3.16.e. Excess flow valve
❏ 3.16.f. Flares

❏ 3.16.g. Scrubbers
❏ 3.16.h. Emergency shutdown systems
❏ 3.16.i. Other (specify) (maximum 200 characters)

3.17. Graphics file name (optional) (maximum 12 characters)

EPA Form 8700-25   -11-

Facility Name: _____

4 | Section 4. Flammables: Worst Case

EPA Facility ID# (leave blank for first submission only)

(If you need to report more than one worst-case scenario, make a photocopy of pages in this section and report each scenario separately)

4.1.a. Chemical Name (maximum 100 characters)

4.2. Model Used (select one or enter another model name in Other below)

- ❑ 4.2.a. EPA's OCA Guidance Reference Tables or Equations
- ❑ 4.2.c. EPA's RMP Guidance for Ammonia Refrigeration Reference Tables or Equations
- ❑ 4.2.d. EPA's RMP Guidance for Waste Water Treatment Plants Reference Tables or Equations
- ❑ 4.2.e. EPA's RMP Guidance for Warehouse Reference Tables or Equations
- ❑ 4.2.f. EPA's RMP Guidance for Chemical Distributors Reference Tables or Equations
- ❑ 4.2.g. EPA's RMP*Comp™
- ❑ 4.2.z. Other model (specify) (maximum 235 characters)

_____

4.3. Scenario (only one option)

Vapor Cloud Explosion

4.4. Quantity released (lbs)	**4.5. Endpoint Used (only one option)**
	1 PSI

4.6. Distance to endpoint (miles)	**4.7. Estimated residential population within distance to endpoint**

4.8. Public Receptors within distance to endpoint (select all that apply)

- ❑ 4.8.a. Schools
- ❑ 4.8.b. Residences
- ❑ 4.8.c Hospitals
- ❑ 4.8.d. Prisons/Correctional facilities
- ❑ 4.8.e. Recreation Areas

- ❑ 4.8.f. Major commercial, office, or industrial areas
- ❑ 4.8.g. Other (specify) (maximum 200 characters)

_____

_____

4.9. Environmental receptors within distance to endpoint (select all that apply)

- ❑ 4.9.a. National or State Parks, Forests, or Monuments
- ❑ 4.9.b. Officially Designated Wildlife Sanctuaries, Preserves, or Refuges
- ❑ 4.9.c. Federal Wilderness Area

- ❑ 4.9.d. Other (specify) (maximum 200 characters)

_____

_____

EPA Form 8700-25   -12-

Facility Name: _____

EPA Facility ID# (leave blank for first submission only)

4   Section 4. Flammables: Worst Case

4.10. Passive mitigation considered (select all that were considered in defining the release quantity or rate for the worst-case scenario)

❑ 4.10.a. Blast walls   ❑ 4.10.b. Other (specify) (maximum 200 characters)

_____

_____

4.11. Graphics file name (optional) (maximum 12 characters)

EPA Form 8700-25   -13-

Facility Name: _____

```
|__|__|__|__| - |__|__|__|__| - |__|__|__|__|
```
EPA Facility ID# (leave blank for first submission only)

5   Section 5. Flammables: Alternative Release

(If you need to report more than one alternative release scenario, make a photocopy of pages in this section and report each scenario separately)

5.1. Chemical Name (maximum 100 characters)

5.2. Model Used (select one or enter another model name in Other below)

❑ 5.2.a. EPA's OCA Guidance Reference Tables or Equations
❑ 5.2.c. EPA's RMP Guidance for Propane Storage Reference Tables or Equations
❑ 5.2.d. EPA's RMP Guidance for Waste Water Treatment Plants Reference Tables or Equations
❑ 5.2.e. EPA's RMP Guidance for Warehouse Reference Tables or Equations
❑ 5.2.f. EPA's RMP Guidance for Chemical Distributors Reference Tables or Equations
❑ 5.2.g. EPA's RMP*Comp™
❑ 5.2.z. Other model (specify) (maximum 235 characters)

_____

5.3. Scenario (select one)

❑ 5.3.a. Vapor cloud explosion   ❑ 5.3.f. Vapor cloud fire
❑ 5.3.b. Fireball   ❑ 3.4.g. Other (specify) (maximum 30 characters)
❑ 5.3.c. BLEVE
❑ 5.3.d. Pool fire   _____
❑ 5.3.e. Jet fire
                                         _____

5.4. Quantity released (lbs)

```
|__|__|__|__|__|__|__|__|__|__|
```

5.5 Endpoint used (select one)

❑ 5.5.a. 1 PSI
❑ 5.5.b. 5 kw/m² for 40 seconds

❑ 5.5.c. Lower flammability limit (specify percent volume) |__|__| . |__|

5.6. Distance to endpoint (miles)	5.7. Estimated residential population within distance to endpoint
|__|__|__| . |__|__|	|__|__| , |__|__|__| , |__|__|__|

RISK MANAGEMENT PLAN – RMP ESUBMIT USER'S MANUAL   *MARCH 2014*

Facility Name: _____

EPA Facility ID# (leave blank for first submission only)

5 Section 5. Flammables: Alternative Release

5.8. Public Receptors within distance to endpoint (select all that apply)

❑ 5.8.a. Schools
❑ 5.8.b. Residences
❑ 5.8.c Hospitals
❑ 5.8.d. Prisons/Correctional facilities
❑ 5.8.e. Recreation Areas

❑ 5.8.f. Major commercial, office, or industrial areas
❑ 5.8.g. Other (specify) (maximum 200 characters)

_____

_____

5.9. Environmental receptors within distance to endpoint (select all that apply)

❑ 5.9.a. National or State Parks, Forests, or Monuments
❑ 5.9.b. Officially Designated Wildlife Sanctuaries, Preserves, or Refuges
❑ 5.9.c. Federal Wilderness Area

❑ 5.9.d. Other (specify) (maximum 200 characters)

_____

_____

5.10. Passive mitigation considered (select all that apply)

❑ 5.10.a. Dikes
❑ 5.10.b. Fire walls
❑ 5.10.c. Blast walls
❑ 5.10.d. Enclosures

❑ 5.10.e. Other (specify) (maximum 200 characters)

_____

_____

5.11. Active mitigation considered (select all that apply)

❑ 5.11.a. Sprinkler systems
❑ 5.11.b. Deluge systems
❑ 5.11.c. Water curtain
❑ 5.11.d. Excess flow valve

❑ 5.11.e. Other (specify) (maximum 200 characters)

_____

_____

5.12. Graphics file name (optional) (maximum 12 characters)

EPA Form 8700-25   -15-

Facility Name: _____

☐ **6** **Section 6. Five-Year Accident History**

☐☐☐☐☐ - ☐☐☐☐☐ - ☐☐☐☐☐
EPA Facility ID# (leave blank for first submission only)

(If you need to report more than one accident history, make a photocopy of pages in this section and report each scenario separately)

Would you like to certify that your facility *did not* have any reportable accidents in the last 5 years?
☐ Yes; leave the rest of this section blank   ☐ No; fill out this section for each accident

6.1. Date of accident (day, month, and year)	6.2. Time accident began (hours and minutes)
☐☐ ☐☐ ☐☐☐☐ M M D D Y Y Y Y	☐☐ ☐☐ ☐ a.m. H H M M ☐ p.m.

6..3. NAICS code of process involved	6.4. Release duration (hours and minutes)
☐☐☐☐☐☐	☐☐☐ ☐☐ H H H M M

6.5.a.i. Chemical name (maximum 100 characters)	6.5.a.ii. CAS Number	6.5.b. Quantity released (lbs.)	6.5.c. Percent weight of chemical if in a mixture (toxics only)
	☐☐☐☐☐-☐☐☐☐-☐		
	☐☐☐☐☐-☐☐☐☐-☐		
	☐☐☐☐☐-☐☐☐☐-☐		
	☐☐☐☐☐-☐☐☐☐-☐		

6.6. Release event (select at least one)

☐ a. Gas release   ☐ d. Explosion
☐ b. Liquid spills/evaporation   ☐ e. Uncontrolled/Runaway Reaction
☐ c. Fire

6.7. Release Source (select at least one)

☐ a. Storage vessel   ☐ g. Joint
☐ b. Piping   ☐ h. Other (specify) (maximum 200 characters)
☐ c. Process vessel
☐ d. Transfer hose   _____
☐ e. Valve
☐ f. Pump   _____

EPA Form 8700-25   -16-

Facility Name: _____

6 **Section 6. Five-Year Accident History**

EPA Facility ID# (leave blank for first submission only)

☐☐☐☐☐ - ☐☐☐☐☐ - ☐☐☐☐☐

6.8. Weather conditions at time of event

a.i. Wind speed (numerical) ☐☐☐ . ☐	Wind speed unit ☐ miles/hr. ☐ knots ☐ meters/sec.	a.ii. Wind direction ☐☐☐
b. Temperature (°F) ☐☐☐	c. Atmospheric stability class (A-F)	☐ d. Precipitation present
☐ e. Unknown weather conditions (check if a-d are all unknown)		

6.9 On-site Impacts

a. Deaths (enter numbers)	b. Injuries (enter numbers)
a.i. Employees or contractors ☐☐☐☐☐	b.i. Employees or contractors ☐☐☐☐☐
a.ii. Public responders ☐☐☐	b.ii. Public responders ☐☐☐
a.iii. Public ☐☐☐☐☐	b.iii. Public ☐☐☐☐☐

c. Property damage

$ ☐☐☐ , ☐☐☐ , ☐☐☐

6.10. Known off-site impacts (enter numbers)

a. Deaths ☐☐☐☐☐☐☐	d. Evacuated ☐☐☐☐☐☐☐
b. Hospitalizations ☐☐☐☐☐☐☐	e. Sheltered-in-place ☐☐☐☐☐☐☐
c. Other medical treatments ☐☐☐☐☐☐☐	f. Property damage ($)
	☐☐☐☐☐☐☐

6.10.g. Environmental damage (select all that apply)

☐ g.1. Fish or animal kills
☐ g.2. Tree, lawn, shrub, or crop damage
☐ g.3. Water contamination
☐ g.4. Soil contamination
☐ g.5. Other (specify) (maximum 200 characters)

_____

_____

EPA Form 8700-25   -17-

Facility Name: _____

| 6 | **Section 6. Five-Year Accident History** |

EPA Facility ID# (leave blank for first submission only)

_____ - _____ - _____

6.11. Initiating event (select one)

❑ a. Equipment failure
❑ b. Human error

❑ c. Natural (weather conditions, earthquake)
❑ d. Unknown

6.12. Contributing factors (select all that apply)

❑ a. Equipment failure
❑ b. Human error
❑ c. Improper procedure
❑ d. Over pressurization
❑ e. Upset condition
❑ f. By-pass condition
❑ g. Maintenance activity/inactivity
❑ h. Process design failure

❑ i. Unsuitable equipment
❑ j. Unusual weather conditions
❑ k. Management error
❑ l. uncontrolled/runaway reaction
❑ m. Other (specify) (maximum 200 characters)

_____

_____

6.13. Off-site responders notified (select one)

❑ a. Notified only
❑ b. Notified and responded

❑ c. No, not notified
❑ d. Unknown

6.14. Changes introduced as a result of the accident (select at least one)

❑ j. None
❑ k. Other (specify) (maximum 200 characters)

❑ a. Improved/upgraded equipment
❑ b. Revised maintenance
❑ c. Revised training
❑ d. Revised operating procedures
❑ e. New process controls
❑ f. New mitigation systems
❑ g. Revised emergency response plan
❑ h. Changed process
❑ i. Reduced inventory

_____

_____

EPA Form 8700-25   -18-

Facility Name: _____

7	**Section 7. Prevention Program: Program 3** EPA Facility ID# (leave blank for first submission only)

(If you need to report more than one prevention program, make a photocopy of pages in this section and report each scenario separately)

Prevention Program description:
_____
_____
_____
_____
_____
_____
_____
_____
_____
_____
_____

7.1. NAICS code for process	
7.2. Chemical name(s) (maximum 100 characters)	

If you need more space to list chemicals, please make a photo copy of this sheet.

7.3. Date on which the safety information was last reviewed or revised

M M   D D   Y Y Y Y

7.4. Process Hazards Analysis (PHA)

7.4.a. Date of last PHA or PHA update

M M   D D   Y Y Y Y

7.4.b. Technique used (select at least one)

❑ 7.4.b.1. What if
❑ 7.4.b.2. Checklist
❑ 7.4.b.3. What if/Checklist combined
❑ 7.4.b.4. HAZOP
❑ 7.4.b.5. Failure Mode & Effects Analysis

❑ 7.4.b.6. Fault Tree Analysis
❑ 7.4.b.7. Other (specify) (maximum 200 characters)

_____

_____

EPA Form 8700-25   -19-

Facility Name: _____

7 Section 7. Prevention Program: Program 3 EPA Facility ID# (leave blank for first submission only)

⌐___ ___ ___ ___⌐ - ⌐___ ___ ___ ___⌐ - ⌐___ ___ ___⌐

7.4.c. Expected or actual date of completion of all changes resulting from last PHA or PHA update

⌐___ ___⌐ ⌐___ ___⌐ ⌐___ ___ ___ ___⌐
M M   D D   Y Y Y Y

7.4.d. Major hazards identified (select at least one)

❏ 7.4.d.1. Toxic release
❏ 7.4.d.2. Fire
❏ 7.4.d.3. Explosion
❏ 7.4.d.4. Runaway reaction
❏ 7.4.d.5. Polymerization
❏ 7.4.d.6. Over pressurization
❏ 7.4.d.7. Corrosion
❏ 7.4.d.8. Overfilling
❏ 7.4.d.9. Contamination

❏ 7.4.d.10. Equipment failure
❏ 7.4.d.11. Loss of cooling, heating, electricity, Instrument air
❏ 7.4.d.12. Earthquake
❏ 7.4.d.13. Floods (flood pain)
❏ 7.4.d.14. Tornado
❏ 7.4.d.15. Hurricanes
❏ 7.4.d.16. Other (specify) (maximum 200 characters)

_____

_____

7.4.e. Process controls in use (select at least one)

❏ 7.4.e.1. Vents
❏ 7.4.e.2. Relief valves
❏ 7.4.e.3. Check valves
❏ 7.4.e.4. Scrubbers
❏ 7.4.e.5. Flares
❏ 7.4.e.6. Manual shutoffs
❏ 7.4.e.7. Automatic shutoffs
❏ 7.4.e.8. Interlocks
❏ 7.4.e.9. Alarms and procedures
❏ 7.4.e.10. Keyed bypass
❏ 7.4.e.11. Emergency air supply

❏ 7.4.e.12. Emergency power
❏ 7.4.e.13. Backup pump
❏ 7.4.e.14. Grounding equipment
❏ 7.4.e.15. Inhibitor addition
❏ 7.4.e.16. Rupture disks
❏ 7.4.e.17. Excess flow device
❏ 7.4.e.18. Quench system
❏ 7.4.e.19. Purge system
❏ 7.4.e.20. None
❏ 7.4.e.21. Other (specify) (maximum 200 characters)

_____

_____

7.4.f. Mitigation systems in use (select at least one)

❏ 7.4.f.1.   Sprinkler system
❏ 7.4.f.2.   Dikes
❏ 7.4.f.3.   Fire walls
❏ 7.4.f.4.   Blast walls
❏ 7.4.f.5.   Deluge system
❏ 7.4.f.6.   Water curtain

❏ 7.4.f.7. Enclosure
❏ 7.4.f.8. Neutralization
❏ 7.4.f.9. None
❏ 7.4.f.10. Other (specify)(maximum 200 characters)

_____

_____

7.4.g. Monitoring/detection systems in use (select at least one)

❏ 7.4.g.1. Process area detectors
❏ 7.4.g.2. Perimeter monitors
❏ 7.4.g.3. None

❏ 7.4.g.4. Other (specify)(maximum 200 characters)

_____

_____

EPA Form 8700-25   -20-

Facility Name: _____

☐ — ☐ — ☐

7 **Section 7. Prevention Program: Program 3**   EPA Facility ID# (leave blank for first submission only)

7.4.h. Changes since last PHA update (select at least one)

☐ 7.4.h.1. Reduction in chemical inventory
☐ 7.4.h.2. Increase in chemical inventory
☐ 7.4.h.3. Change in process parameters
☐ 7.4.h.4. Installation of process controls
☐ 7.4.h.5. Installation of process detection systems
☐ 7.4.h.6. Installation of perimeter monitoring systems
☐ 7.4.h.7. Installation of mitigation systems

☐ 7.4.h.8.   None recommended
☐ 7.4.h.9.   None
☐ 7.4.h.10. Other (specify) (maximum 200 characters)

_____

_____

7.5. Date of most recent review or revision of operating procedures

☐☐ ☐☐ ☐☐☐☐
M M D D Y Y Y Y

7.6. Training

7.6.a. Date of most recent review or review of operating procedures

☐☐ ☐☐ ☐☐☐☐
M M D D Y Y Y Y

7.6.b. Type of training provided (select at one)

☐ 7.6.b.1. Classroom
☐ 7.6.b.2. On the job
☐ 7.6.b.3. Other (specify) (maximum 200 characters) _____

7.6.c. Type of competency testing used (select at least one)

☐ 7.6.c.1. Written test
☐ 7.6.c.2. Oral Test
☐ 7.6.c.3. Demonstration

☐ 7.6.c.4. Observation
☐ 7.6.c.5. Other (specify)(maximum 200 characters)

_____

_____

7.7. Maintenance

7.7.a. Date of most recent review or revision of maintenance procedures

☐☐ ☐☐ ☐☐☐☐
M M D D Y Y Y Y

7.7.b. Date of most recent equipment inspection or test

☐☐ ☐☐ ☐☐☐☐
M M D D Y Y Y Y

7.7.c. Equipment most recently inspected or tested (list equipment) (maximum 200 characters)

EPA Form 8700-25   -21-

Facility Name: _____

| 7 | Section 7. Prevention Program: Program 3 | EPA Facility ID# (leave blank for first submission only) |

_____ - _____ - _____

7.8 Management of Change

7.8.a. Date of most recent changes that triggered management of change procedures.	M M D D Y Y Y Y
7.8.b. Date of most recent changes that triggered management of change procedures.	M M D D Y Y Y Y
7.9. Date of most recent pre-startup review	M M D D Y Y Y Y

7.10. Compliance audits

7.10.a. Date of most recent compliant audit	M M D D Y Y Y Y
7.10.b. Expected or actual date of completion of all changes resulting from the compliance audit	M M D D Y Y Y Y

7.11. Incident investigation

7.11.a. Date of most recent incident investigation (if any)	M M D D Y Y Y Y
7.11.b. Expected or actual date of completion of all changes resulting from the incident investigation	M M D D Y Y Y Y

7.12. Date of most recent review or revision of employee participation plans	M M D D Y Y Y Y

7.13. Date of most recent review or revision of hot work permit procedures	M M D D Y Y Y Y

7.14. Date of most recent review or revision of contractor safety procedures	M M D D Y Y Y Y

7.15. Date of most recent review or revision of contractor safety performance	M M D D Y Y Y Y

EPA Form 8700-25   -22-

Facility Name: _____

8 **Section 8. Prevention Program: Program 2**   EPA Facility ID# (leave blank for first submission only)

└─┴─┴─┘ - └─┴─┴─┘ - └─┴─┴─┴─┘

(If you need to report more than one prevention program, make a photocopy of pages in this section and report each scenario separately)

Prevention Program description:

_____
_____
_____
_____
_____
_____
_____
_____
_____
_____
_____

8.1. NAICS code for process	└─┴─┴─┴─┴─┴─┘
8.2. Chemical name(s) (maximum 100 characters)	

If you need more space to list chemicals, please make a photo copy of this sheet.

8.3 Safety Information

8.3. Date of most recent review or revision of safety information	└─┴─┘ └─┴─┘ └─┴─┴─┴─┘
	M M D D Y Y Y Y

8.3.b. Federal/state regulations or industry-specific design codes and standards used to demonstrate compliance with safety information requirement (select at least one)

❏ 8.3.b.1. NFPA 58 (or state law based on NFPA 58)
❏ 8.3.b.2. OSHA (29 CFR 1910.111)
❏ 8.3.b.3. ASTM Standards
❏ 8.3.b.4. ANSI Standards
❏ 8.3.b.5. ANSME Standards
❏ 8.3.b.6. None
❏ 8.3.b.8. Comments (100 characters)

❏ 8.3..b.7. Other (specify) (maximum 200 characters)

_____

_____

_____

_____

EPA Form 8700-25   -23-

Facility Name: _____

<table>
<tr><td>**8**</td><td>**Section 8. Prevention Program: Program 2**</td><td>⌐ ⌐ ⌐ - ⌐ ⌐ ⌐ - ⌐ ⌐ ⌐
EPA Facility ID# (leave blank for first submission only)</td></tr>
</table>

8.4. Hazard review

8.4.a. Date of completion of most recent hazard review or update	⌐⌐ ⌐⌐ ⌐⌐⌐⌐ M M D D Y Y Y Y
8.4.b. Expected or actual date of completion of all changes resulting from the hazard review	⌐⌐ ⌐⌐ ⌐⌐⌐⌐ M M D D Y Y Y Y

8.4.c. Major hazards identified (select at least one)

- ❑ 8.4.c.1. Toxic release
- ❑ 8.4.c.2. Fire
- ❑ 8.4.c.3. Explosion
- ❑ 8.4.c.4. Runaway reaction
- ❑ 8.4.c.5. Polymerization
- ❑ 8.4.c.6. Over pressurization
- ❑ 8.4.c.7. Corrosion
- ❑ 8.4.c.8. Overfilling
- ❑ 8.4.c.9. Contamination
- ❑ 8.4.c.10. Equipment failure

- ❑ 8.4.c.11. Loss of cooling, heating, electricity, instrument air
- ❑ 8.4.c.12. Earthquake
- ❑ 8.4.c.13. Floods (flood pain)
- ❑ 8.4.c.14. Tornado
- ❑ 8.4.c.15. Hurricanes
- ❑ 8.4.c.16. Other (specify) (maximum 200 characters)

_____

_____

8.4.d. Process controls in use (select at least one)

- ❑ 8.4.d.1. Vents
- ❑ 8.4.d.2. Relief valves
- ❑ 8.4.d.3. Check valves
- ❑ 8.4.d.4. Scrubbers
- ❑ 8.4.d.5. Flares
- ❑ 8.4.d.6. Manual shutoffs
- ❑ 8.4.d.7. Automatic shutoffs
- ❑ 8.4.d.8. Interlocks
- ❑ 8.4.d.9. Alarms and procedures
- ❑ 8.4.d.10. Keyed bypass
- ❑ 8.4.d.11. Emergency air supply
- ❑ 8.4.d.12. Emergency power

- ❑ 8.4.d.13. Backup pump
- ❑ 8.4.d.14. Grounding equipment
- ❑ 8.4.d.15. Inhibitor addition
- ❑ 8.4.d.16. Rupture disks
- ❑ 8.4.d.17. Excess flow device
- ❑ 8.4.d.18. Quench system
- ❑ 8.4.d.19. Purge system
- ❑ 8.4.d.20. None
- ❑ 8.4.d.21. Other (specify) (maximum 200 characters)

_____

_____

EPA Form 8700-25   -24-

Facility Name: _____

⌞⌟⌞⌟ - ⌞⌟⌞⌟ - ⌞⌟⌞⌟

EPA Facility ID# (leave blank for first submission only)

8 **Section 8. Prevention Program: Program 2**

8.4.e. Mitigation systems in use (select at least one)
- ❏ 8.4.e.1. Sprinkler system
- ❏ 8.4.e.2. Dikes
- ❏ 8.4.e.3. Fire walls
- ❏ 8.4.e.4. Blast walls
- ❏ 8.4.e.5. Deluge system
- ❏ 8.4.e.6. Water curtain
- ❏ 8.4.e.7. Enclosure

- ❏ 8.4.e.8. Neutralization
- ❏ 8.4.e.9. None
- ❏ 8.4.e.10. Other (specify)(maximum 200 characters)

_____

_____

8.4.f. Monitoring/detection systems in use (select at least one)

- ❏ 8.4.f.1. Process area detectors
- ❏ 8.4.f.2. Perimeter monitors
- ❏ 8.4.f.3. None

- ❏ 8.4.f.4 . Other (specify)(maximum 200 characters)

_____

_____

8.4.g. Changes since last hazard review or hazard review update (select at least one)

- ❏ 8.4.g.1. Reduction in chemical inventory
- ❏ 8.4.g.2. Increase in chemical inventory
- ❏ 8.4.g.3. Change in process parameters
- ❏ 8.4.g.4. Installation of process controls
- ❏ 8.4.g.5. Installation of process detection systems
- ❏ 8.4.g.6. Installation of perimeter monitoring systems
- ❏ 8.4.g.7. Installation of mitigation systems

- ❏ 8.4.g.8. None recommended
- ❏ 8.4.g.9. None
- ❏ 8.4.g.10. Other (specify) (maximum 200 characters)

_____

_____

8.5. Date of most recent review or revision of safety information   ⌞⌟⌞⌟   ⌞⌟⌞⌟   ⌞⌟⌞⌟⌞⌟
                                                                          M M   D D   Y Y Y Y

8.6. Training

8.6.a. Date of most recent review or revision of training programs   ⌞⌟⌞⌟   ⌞⌟⌞⌟   ⌞⌟⌞⌟⌞⌟
                                                                        M M   D D   Y Y Y Y

8.6.b. Type of training provided (select at one)

- ❏ 8.6.b.1. Classroom
- ❏ 8.6.b.2. On the job
- ❏ 8.6.b.3. Other (specify) (maximum 200 characters) _____

_____

EPA Form 8700-25   -25-

Facility Name: _____

⌷⌷⌷⌷ - ⌷⌷⌷⌷ - ⌷⌷⌷⌷

EPA Facility ID# (leave blank for first submission only)

8 **Section 8. Prevention Program: Program 2**

8.6.c. Type of competency testing used (select at least one)   ❑ 8.6.c.5. Other (specify)(maximum 200 characters)

❑ 8.6.c.1. Written test
❑ 8.6.c.2. Oral Test   _____
❑ 8.6.c.3. Demonstration
❑ 8.6.c.4. Observation   _____

8.7. Maintenance

8.7.a. Date of most recent review or revision of maintenance procedures	⌷⌷ ⌷⌷ ⌷⌷⌷⌷ M M D D Y Y Y Y
8.7.b. Date of most recent equipment inspection or test	⌷⌷ ⌷⌷ ⌷⌷⌷⌷ M M D D Y Y Y Y

8.7.c. Equipment most recently inspected or tested (list equipment) (maximum 200 characters)

8.8. Compliance audits

8.8.a. Date of most recent compliant audit	⌷⌷ ⌷⌷ ⌷⌷⌷⌷ M M D D Y Y Y Y
8.8.b. Expected or actual date of completion of all changes resulting from the compliance audit	⌷⌷ ⌷⌷ ⌷⌷⌷⌷ M M D D Y Y Y Y

8.9. Incident investigation

8.9.a. Date of most recent incident investigation (if any)	⌷⌷ ⌷⌷ ⌷⌷⌷⌷ M M D D Y Y Y Y
8.9.b. Expected or actual date of completion of all changes resulting from the incident investigation	⌷⌷ ⌷⌷ ⌷⌷⌷⌷ M M D D Y Y Y Y

8.10. Date of most recent change that triggered a review or a revision of safety information, the hazard review, operating or maintenance procedures, or training	⌷⌷ ⌷⌷ ⌷⌷⌷⌷ M M D D Y Y Y Y

EPA Form 8700-25   -26-

Facility Name: _____

                                                    ▭▭▭▭▭ - ▭▭▭▭▭ - ▭▭▭▭▭
                                                    EPA Facility ID# (leave blank for first submission only)

| 9 | **Section 9. Emergency Response** |

9.1 Written emergency response (ER) plan

9.1.a. ❏ Is your facility included in the written community emergency response plan?

9.1.b. ❏ Does your facility have its own written emergency response plan?

9.2. ❏ Does your facility's ER plan include specific actions to be taken in response to accidental releases of regulated substance(s)?

9.3. ❏ Does your facility's ER plan include procedures for informing the public and local agencies responding to accidental releases?

9.4. ❏ Does your facility's ER plan include information on emergency health care?

9.5. Date of most recent review or update of your facility's ER plan

                                                  ▭▭   ▭▭   ▭▭▭▭
     M M   D D   Y Y Y Y

9.6. Date of most recent ER training for your facility's employees

                                  ▭▭   ▭▭   ▭▭▭▭
     M M   D D   Y Y Y Y

9.7. Local agency with which your facility's ER plan or response activities are coordinated

9.7.a. Name of agency (maximum 35 characters)

9.7.b. Phone number (▭▭▭) ▭▭▭ - ▭▭▭▭

9.8. Subject to (select all that apply)

❏ 9.8.a. OSHA Regulations at 29 CFR 1910.38
❏ 9.8.b. OSHA Regulations at 29 CFR 1910.120
❏ 9.8.c. Clean Water Act Regulations at 40 CFR 112
❏ 9.8.d. RCRA Regulations at 40 CFR 264, 265, 279.52
❏ 9.8.e. OPA-90 Regulations at 40 CFR 112, 33 CFR 154, 49 CFR 194, 30 CFR 254
❏ 9.8.f. State EPCRA Rules or Laws
❏ 9.8.g. Other (specify)(maximum 200 characters)

_____

_____

EPA Form 8700-25   -27-   Return to Index

APPENDIX B. SAMPLE CERTIFICATION LETTERS

SAMPLE CERTIFICATION LETTERS

Certification Statement for Program 1 Process(es):

Based on the criteria in 40 CFR 68.10, the distance to the specified endpoint for the worst-case accidental release scenario for the following process(es) is less than the distance to the nearest public receptor:

- [insert description for first program 1 process from executive summary]

- [insert description for second program 1 process from executive summary]

- etc.

Within the past five years, the process(es) has (have) had no accidental release that caused offsite impacts provided in the risk management program rule (40 CFR 68.10(b)(1)). No additional measures are necessary to prevent offsite impacts from accidental releases. In the event of fire, explosion, or a release of a regulated substance from the process(es), entry within the distance to the specified endpoints may pose a danger to public emergency responders. Therefore, public emergency responders should not enter this area except as arranged with the emergency contact indicated in the RMP. The undersigned certifies that, to the best of my knowledge, information, and belief, formed after reasonable inquiry, the information submitted is true, accurate, and complete.

_____   _____
Signature   Print Name

_____   _____
Title   Date

Certification Statement for Program Level 2 & 3 Processes:

To the best of the undersigned's knowledge, information, and belief formed after reasonable inquiry, the information submitted is true, accurate, and complete.

_____   _____
Signature   Print Name

_____   _____
Title   Date

B-1

<u>Certification Statement for a Correction:</u>

To the best of the undersigned's knowledge, information, and belief formed after reasonable inquiry, these corrections and/or administrative changes are true, accurate, and complete.

_____   _____
Signature   Print Name

_____   _____
Title   Date

EPA Facility ID # ☐☐☐☐ - ☐☐☐☐ - ☐☐☐☐

APPENDIX C. CBI SUBSTANTIATION FORM

OM B Control Number: 2050-0144

CBI SUBSTANTIATION FORM

If you are claiming Confidential Business Information (CBI) in your Risk Management plan (RMP), you must substantiate your claim at the same time that you submit your RMP. To qualify for CBI protection, the substantive criteria in 40 CFR 2.301 must be met. Certain RMP data elements cannot be claimed CBI, as stated in 40 CFR 68.151.

Fill out this form for each data element or set of data elements that have a discrete substantiation. You may use one CBI Substantiation Form to report multiple data elements as CBI if the basis for substantiation is the same. That means the answers to the questions in Part IV must be the same for all the data elements. If you need more space in Part III, please attach a separate piece of paper.

Burden Statement

The public reporting and recordkeeping burden for this collection of information is estimated to average 8.5 hours per claim. Burden means the total time, effort, or financial resources expended by persons to generate, maintain, retain, or disclose or provide information to or for a Federal agency. This includes the time needed to review instructions; develop, acquire, install, and utilize technology and systems for the purposes of collecting, validating, and verifying information, processing and maintaining information, and disclosing and providing information; adjust the existing requirements; train personnel to be able to respond to a collection of information; search data sources; complete and review the collection of information; and transmit or otherwise disclose the information. An agency may not conduct or sponsor, and a person is not required to respond to, a collection of information unless it displays a currently valid OMB control number.

Send comments on the Agency's need for this information, the accuracy of the provided burden estimates, and any suggested methods for minimizing respondent burden, including through the use of automated collection techniques to the Director, OPPE Regulatory Information Division, U.S. Environmental Protection Agency (2137), 401 M St., S.W., Washington D.C. 20460. Include the OMB control number in any correspondence. Do not send the completed CBI substantiation to this address.

Part I -- Facility Identification Information

The information given here must correspond to the information that you provided in the registration section of your RMP. If you have an EPA Facility ID #, please include this information. If you are resubmitting, updating or correcting your RMP, you should already have received an EPA Facility ID#.

a. Facility Name:
b. EPA Facility ID # (if assigned): ☐☐☐☐ - ☐☐☐☐ - ☐☐☐☐
c. Facility Location Address:
d. City, State and Zip Code:

EPA Form 8700-27   C-1

e. Dun and Bradstreet Number:	

Part II – Is this substantiation a sanitized or an unsanitized version?
If this substantiation contains any CBI, you must also submit a sanitized substantiation (without CBI data) as stated in 40 CFR 68.152. In this case, submit 2 copies of this form, one sanitized and one unsanitized. Please indicate here whether this form is sanitized or unsanitized.

☐   Sanitized   ☐   Unsanitized

Part III – List the RMP Data Elements which you are claiming CBI that are covered in this substantiation form. List the data element number and its descriptive name, but NOT the actual CBI data. Please note that you may use one substantiation form for more than one data element only if the answers to all of the questions in Part IV are the same for those data elements.

Data Element #	Data Element Name

EPA Form 8700-27   C-2

Part IV – The following are criteria set forth in 40 CFR 2.204, 2.208 and 2.301 for substanti ating CBI claims. Provide answers to each of the following questions to substantiate your claim. If you need additional space, use separate sheets of paper.

(a)   For any data elements that you wish to claim CBI that are listed in Part III, please indicate whether your business has previously submitted a CBI claim for this data element to EPA and whether that claim has expired, been waived, or been withdrawn.

(b)   What reasonable measures have you taken to protect the confidentiality of the information and do you intend to continue to take these measures?

EPA Form 8700-27   C-3

(c)	Have you disclosed the information to anyone other than a governmental body? If so, why should the information still be considered confidential? If not, is the information reasonably obtainable without your consent? Has EPA or another Federal agency made a determination as to the confidentiality of the information? If so, please attach a copy of the determination.
(d)	Does any statute require public disclosure of the information for which you are claiming CBI? If so, identify the law.

EPA Form 8700-27   C-4

(e)	(1) For each data element claimed as CBI in Part III, discuss with specificity why release of the information is likely to cause substantial harm to your competitive position. Explain the nature of those harmful effects, why they should be viewed as substantial, and the causal relationship between disclosure and such harmful effects. For example, how could your competitors make use of this information to your detriment?
	(2) Do you assert that the information is "voluntarily submitted" as defined at 40 CFR 2.201(i)? If so, explain why, and how disclosure would tend to lessen the Governments's ability to obtain necessary information in the future.

Part V - Certification (Read and sign after completing all sections)
To the best of the undersigned's knowledge, information, and belief formed after reasonable inquiry, the information submitted is true, accurate, and complete.

Name and official title of owner or operator or senior management official

Signature (All signatures must be original)	Print Name
Official Title	Date Signed

APPENDIX D. CBI UNSANITIZED DATA ELEMENT FORM

OMB Control Number: 2050-0144

CBI UNSANITIZED DATA ELEMENT FORM

If you are claiming Confidential Business Information (CBI) in your RMP, you must submit in paper form both the information being claimed CBI and a substantiation for your claim at the time you submit your redacted or "sanitized" RMP . This form should be used to submit the confidential information. The redacted RMP will be made available to the public in RMP*Info.

If you need additional space, make a copy of page 2 of this form.

Part I. Facility Identification Information
The information given here should correspond to the information that you filled out in the registration section of your RMP. If you have an EPA Facili ty ID#, please include this information. You will have received the number after your first submission.

a. Facility Name:	
b. EPA Facility ID # (if assigned):	☐☐☐☐ - ☐☐☐☐ - ☐☐☐☐
c. Facility Location Address:	
d. City, State and Zip Code:	
e. Dun and Bradstreet Number:	

Part II - Information claimed as CBI
Please list the data element number(s) from the RMP form (paper form or electronic form), the name(s) of the element(s) you are claiming CBI, and the actual CBI data.

Data Element Number	Name of Data Element	RMP Data Claimed as CBI

EPA Form 8700-28   D-1

Data Element Number	Name of Data Element	RMP Data Claimed as CBI

APPENDIX E. RISK MANAGEMENT PROGRAM DE-REGISTRATION FORM

Today's Date: _____

EPA Facility Identifier: _____

Effective Date of De-registration: _____

Facility Name: _____

Facility Address: _____

_____

City: _____**State:** _____**Zip Code:** _____

Select (Check) Reason for De-registration:

       □ Source reduced inventory of all regulated substances below TQs
       □ Source no longer uses any regulated substance
       □ Source terminated operations
       □ Other: _____

I, _____, certify the above stationary source as of the above
         (Name of Facility Owner or Operator)
effective date is no longer covered by the Accidental Release Prevention Regulations, 40 CFR Part 68.

_____   _____
   Signature of Owner or Operator   Date

_____
        Official Title

PLEASE MAIL THE COMPLETED DE-REGISTRATION FORM PROMPTLY TO:

    U.S. Environmental Protection Agency
    Attention: RMP Reporting Center
    P.O Box 10162
    Fairfax, VA 22038

If you prefer to send your De-registration Form by certified mail, courier or overnight mail (e.g., Fed Ex, UPS, etc.), please address it to:

    RMP Reporting Center
    c/o CGI Federal, Inc.
    12601 Fair Lakes Circle
    Fairfax, VA 22033

APPENDIX F. RISK MANAGEMENT PROGRAM PROPANE WITHDRAWAL FORM

EPA Facility Identification #: _____

Facility Name: _____

Facility Location Address: _____

_____

City: _____ State: _____ Zip Code: _____

The facility listed above is withdrawing its RMP submission per 40 CFR 68.126 because the facility does not have more than a threshold quantity of any regulated substance listed at 40 CFR 68.130 other than propane (or another listed flammable substance):

 ☐ that the facility uses on site as a fuel,

  or

 ☐ that the facility holds for retail sale as fuel. More than one-half of the income of this facility is obtained from direct sales of the fuel to end users or more than one-half of the fuel sold, by volume, is sold through a cylinder exchange program.

_____ _____ _____
  Operator/Owner Name   Official Title   Date

PLEASE MAIL THIS COMPLETED PROPANE WITHDRAWAL FORM PROMPTLY TO:

 U.S. Environmental Protection Agency
 Attention: RMP Reporting Center
 P.O Box 10162
 Fairfax, VA 22038

If you prefer to send this Propane Withdrawal Form by certified mail, courier or overnight mail (e.g., Fed Ex, UPS, etc.), please address it to:

 RMP Reporting Center
 c/o CGI Federal, Inc.
 12601 Fair Lakes Circle
 Fairfax, VA 22033

APPENDIX G. RISK MANAGEMENT PROGRAM REQUEST TO CONSOLIDATE EPA FACILITY ID NUMBERS

EPA Facility Identification #: _____

EPA Facility Identification #: _____

EPA Facility Identification #: _____

Facility Name: _____

Facility Location Address: _____

City: _____ **State:** _____ **Zip Code:** _____

The facility listed above has been assigned more than one EPA Facility ID Number. This form requests the EPA Facility ID Numbers to be consolidated to a single ID Number. Please de-activate additional/incorrect Facility IDs Numbers.

_____	_____	_____
Operator/Owner Name	Official Title	Date

PLEASE MAIL THIS REQUEST TO CONSOLIDATE EPA FACILITY ID NUMBERS FORM PROMPTLY TO:

> **U.S. Environmental Protection Agency**
> **Attention: RMP Reporting Center**
> **P.O. Box 10162**
> **Fairfax, VA 22038**

If you prefer to send this Request to Consolidate EPA Facility ID Numbers Form by certified mail, courier or overnight mail (e.g., Fed Ex, UPS, etc.), please address it to:

> RMP Reporting Center
> c/o CGI Federal, Inc.
> 12601 Fair Lakes Circle
> Fairfax, VA 22033